MW01516046

Adhesive Bonding in Photonics Assembly and Packaging

Adhesive Bonding in Photonics Assembly and Packaging

B. G. Yacobi

University of Toronto,
Department of Materials Science and Engineering,
Toronto, Ontario, Canada

and

M. Hubert

EXFO Photonic Solutions Inc.,
Mississauga, Ontario, Canada

AMERICAN SCIENTIFIC PUBLISHERS
25650 North Lewis Way
Stevenson Ranch, California 91381-1439, USA

AMERICAN SCIENTIFIC PUBLISHERS

25650 North Lewis Way, Stevenson Ranch, California 91381-1439, USA
Tel.: (661) 254-0807, Fax: (661) 254-1207
E-mail: order@aspbs.com
URL: http://www.aspbs.com

Adhesive Bonding in Photonics Assembly and Packaging by B. G. Yacobi and M. Hubert.

This book is printed on acid free paper. ∞

Library of Congress Catalog Number: 2003106910
International Standard Book Number: 1-58883-019-5

PRINTED IN THE UNITED STATES OF AMERICA
10 9 8 7 6 5 4 3 2 1

Preface

Adhesives are increasingly being used in the manufacture of optoelectronic and fiber-optic components and assemblies. It is important to understand their limitations and how to use them optimally. Furthermore, in these applications, optical radiation curing of adhesives is becoming the preferred method of providing high cure speeds and increased throughputs. The main objective of this book is to outline the issues related to optical adhesives and adhesive bonding (e.g., their properties, methods of curing, use, degradation, and reliability) with a general emphasis on their applications to joining various optical components and fiber-optic assemblies.

A critical issue in the manufacture of fiber-optic components and various telecommunications assemblies is the development of automated processes for producing consistent (or identical) parts with increased yields and reduced cost. This issue is further complicated by the wide variation in package designs and the lack of standardized alignment and joining methodologies and a winning technology. The great effort from the manufacturer's perspective is directed toward producing high-volume and low-cost photonics devices and assemblies. Such an efficient manufacturing requires the availability of appropriate processing methods and instrumentation, as well as high-volume assembly and packaging technologies, which would enable the fabrication lines to significantly reduce the cost of production of photonics components and assemblies.

It should be noted that in the description of related issues in the literature, various terms, such as assembly, packaging, and integration, are often employed, which may cause ambiguity. In the context of the present book, we find the terms *assembly* and *packaging* are most suitable, since they encompass various levels of structure or device manufacturing processes related to photonics and microelectronics applications. However, in some cases of integrated manufacturing technology, the term *assembly* is strictly used to describe all processes, for example, alignment, adjustment, and physical attachment of materials and components using various means, including bonding, soldering, and sealing, which are associated with forming the end product; the term *packaging* (in manufacturing technology) is often related to shipping and labeling processes. However, in the context of the present book, the term *packaging* strictly implies a process (e.g., hermetic sealing) for providing a means of protection of the components, assemblies, and devices (as end products) from the environment.

This book has evolved from a review (published in the *Journal of Applied Physics*), which was written by the authors together with K. Davis, A. Hudson, and S. Martin, and we would like to thank them for valuable discussions and their contributions (including figures) to different chapters of this book. We are especially grateful to K. Davis for many useful discussions and suggestions related to the application issues of adhesive bonding. The authors would also like to thank A. Firhoj for helpful suggestions, Dr. D. V. Heyd for his data related to Raman measurements and helpful discussions, and M. Singh for providing illustrations for the book. The authors are grateful to EXFO Corporation for its encouragement and support in writing this book, and to Dr. F. G. Shi for reviewing the manuscript and useful suggestions.

Contents

About the Authors

Dr. B. G. Yacobi is an Adjunct Professor in the Department of Materials Science and Engineering at the University of Toronto. His experience is mainly in the synthesis, applications, and analysis of electronic and optoelectronic materials and devices, with about ninety publications including articles in refereed journals, reviews, patents, book chapters, and books. During his tenure in various academic, industrial, and government institutions in the United States and Canada, his research was related to crystalline and amorphous semiconductors, diamond films, epitaxial heterostructures, nanostructured materials and their characterization.

Dr. M. Hubert is the Vice President Research of EXFO Photonics Solutions, an EXFO company that specializes in providing adhesive bonding solutions for the fiber-optic manufacturing sector and is the leading supplier of UV spot curing systems. Dr. Hubert, a graduate of the University of Toronto, has been involved with photonics for over 30 years. He has done research and published papers in the area of gas lasers, Raman and Brillouin spectroscopy, confocal microscopy, fiber Bragg gratings and photoadhesives. He served for ten years as Director of the Resource Facility at the Ontario Laser and Lightwave Research Center (now Photonics Research Ontario), a provincial Center of Excellence at the University of Toronto.

CHAPTER 1

Introduction

Fiber-optic communication and information processing systems involve the generation, transmission, and control of light. These include the emission, amplification, transmission, and detection of light by employing various optical components and devices. In such systems, optical fibers and components are most commonly joined with ultraviolet (UV)-cured or thermally cured adhesives. Among myriad types of adhesives, optical adhesives, which may be included as part of the optical path, are characterized by a refractive index adapted to the component material of the optical assembly. The on-going demands for high-reliability and low-cost optical assemblies catalyze the development of optical adhesives with refractive index and molecular structure being designed (by optimizing formulations) to provide controllable refractive indices in the range between about 1.40 and 1.60. These are typically ultraviolet-cured optical adhesives with very good matching of the refractive index with optical components of the assembly. The general trend in this context is the development of optical adhesives with advantageous optical, mechanical, and thermal properties, which would facilitate the use of adhesive bonding in joining of various materials with different characteristics.

It should be noted at this juncture that no single adhesive can be selected for a wide variety of applications, and thus, one of the essential tasks is identifying (and even developing, if necessary) suitable adhesives for specific applications.

A critical issue related to optical assemblies is that of reliability, such as the long-term stability of the optical alignment (e.g., between optical fiber and laser), stability and degradation issues related to optical adhesives, and the requirement of airtight sealing. The general trends in the development of improved properties of optical adhesives include (i) greater strength bonds, resulting in improved durability; (ii) faster cure rates; and (iii) lower stress.

A wide variety of optical, electro-optical, and optoelectronic components and devices are employed in fiber-optic communication systems. These include components and devices for the generation, modulation, guiding, amplification, switching, and detection of electromagnetic radiation. Having various sizes and tolerances, these components and devices are assembled into a package (or module) designed to include the coupling of the electromagnetic radiation into and/or out of optical fiber cables.

A schematic diagram illustrating the basic components of the optical fiber communication system is given in Figure 1.1. Digitized forms of information in the binary format (i.e., bits, 0s and 1s) are provided by the *encoder* as an electrical signal is converted into a current and subsequently into optical pulses by the *electrical-to-optical converter* (i.e., a current-driven light source, which is typically a semiconductor laser). Thus, information represented by bytes of data travels through the fiber-optic cable at the speed of light. At another end (i.e., the receiver end), a photodetector (i.e., *optical-to-electrical converter*) detects and converts the optical signal into an electrical signal, and then the *decoder* converts it back to bytes of data. (Note that additional components, such as repeaters and/or optical amplifiers, are also incorporated in the fiber-optic systems.) It should be emphasized that various splices and connectors facilitate the interfacing (i.e., joining) between different components of this system.

As mentioned above, in such optical fibers and components there are many bonds made with adhesives, which provide great flexibility and versatility for joining dissimilar materials. In order not to lose power at the interface, optical adhesives are selected so that their index of refraction is matched to the index of refraction of the optical fiber. When adhesives are used, optical radiation curing of these adhesives is becoming the preferred method providing high cure speeds and throughputs.

One of the critical issues in manufacturing of various optical assemblies is that of automation, which is not a trivial task considering the wide variation in package designs and the lack of standardized alignment and joining methodologies. Furthermore, assembling optical components generally requires very high precision and placement accuracy (of less than 1 μm). In such cases, automated assembly, involving machine-assisted alignment and attachment processes that facilitate large-volume manufacturing, provides possible solutions to these issues.

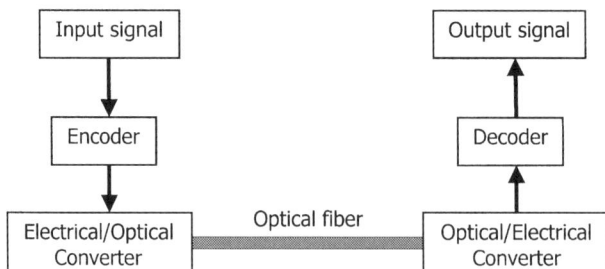

Figure 1.1   Schematic diagram illustrating the basic components of the fiber-optic communication system.

It will be useful to clarify the basic terminology used in the description as it refers to such terms as *optical, electro-optical*, and *optoelectronic*. Optics, in general, refers to the branch of physics related to vision and some phenomena associated with electromagnetic radiation in the wavelength range between the vacuum ultraviolet (about 40 nm) and the far infrared (1 mm). A more comprehensive term that is commonly used in the literature is that of *photonics*, which refers to the technology field related to the generation, transmission, and control of light for communications and information processing (it includes the emission, deflection, amplification, transmission, and detection of light by employing various optical components and devices, such as fiber optics, light sources, and electro-optical and optoelectronic instrumentation). In this context, the term *electro-optics* refers to the field of science related to the use of applied electrical fields to generate and control optical radiation (note that this term is frequently used to describe topics related to optoelectronics). The term *optoelectronics* refers to the technology related to the generation of light (e.g., lasers and light-emitting diodes), amplification of light, control of light, and detection of light.

It should be noted that, in addition to assemblies related to fiber-optic communication systems, optical adhesives are widely employed in mounting such passive components as lenses, prisms, filters, windows, and mirrors in various optical instruments.

Another major application of optical adhesive bonding is in various *optomechanical* systems requiring precision (and secure) mounting of optical components. Optomechanical adhesive bonding has even more stringent requirements than other adhesive bonding applications, since it is essential that, in order to prevent any optical distortion, stress is minimized throughout the operating temperature limits and adhesively bonded components remain precisely positioned during their operating lifetime.

The main objective of this book is to provide a description of the properties, applications, reliability, and degradation issues related to optical adhesives, as well as to provide the reader with some guidelines for their effective use in photonics applications. It should be pointed out that the description in this book is related to the present-day technology of adhesive bonding, employing light-cured adhesives. With continuing developments in both polymer science and photonics applications, many aspects of adhesive science and technology will certainly undergo corresponding advances, some of which are discussed in a later chapter in this book (Chapter 8).

The topics covered in the book are organized in chapters as follows. Chapter 2 presents a brief description of optical and photonics components and assemblies. Next, in Chapters 3 and 4, there is an introduction to the basic principles of adhesive bonding and types of adhesives. These chapters (i.e., Chapters 2 through 4) provide the necessary foundation for further discussions in the following chapters dedicated to the issues of photopolymerization, adhesive curing, applications, and problems related to optical adhesive bonding. Since these topics have major importance in understanding the mechanisms and processes underlying optical adhesive bonding, we devote separate chapters to each of these topics. Chapter 5 provides a description of the processes and methods related to optical radiation curing, as well as the mechanisms of curing and characteristics of cured adhesives, and the issue of monitoring the degree of adhesive curing. Applications of adhesive bonding in assembling various photonics structures and devices are outlined in Chapter 6. The issues and problems related to the reliability and degradation of optical adhesive bonding are outlined in Chapter 7. Finally, future directions of developments are discussed in Chapter 8.

We hope that this discussion will provide sufficient information on adhesive bonding for optical and photonics components and systems, so that the reader will develop sufficient knowledge related to (a) the basic principles and applications of adhesion and adhesives and their classification types; (b) the properties of adhesives that determine their applications; (c) the importance of design considerations related to specific optical assemblies; (d) the importance of issues such as cleanliness and contamination removal; and (e) the stresses, degradation, and reliability issues.

It should be noted that this book is concerned with optical-radiation-induced curing of adhesives. This method for adhesive curing, used in the most optimal manner, is becoming one of the primary methods of choice for bonding of components in fiber-optic assemblies.

It is natural to compare the various practical issues of applications of adhesives in photonics assembly and packaging to those related to electronics assembly and packaging, which is a well-developed and mature technology. Although some of the issues are fairly similar, one of the important differences is related to the fact that, in photonics assembly and packaging, the geometrical alignment is much more stringent and of much greater consequence. Indeed, in the case of electronics applications, in which the functionality of the assembly relates to electron transport, certain paths (or gaps) for electrons can be ensured by incorporating suitable solders, for example. In contrast, in photonics applications (e.g., coupling of two single-mode optical fibers), the geometric and structural considerations are more stringent. Such considerations (i.e., alignment issues), directed at preventing, or, in practical terms, minimizing any transmission losses of light at the joint, are discussed in the following chapter.

It is necessary to emphasize the importance of efficient manufacturing practices for photonics assembly and packaging. Presently, the major cost of manufacturing of photonics components is in the assembly and packaging, related to such steps as the mutual alignment of optical elements and of the optical fiber; such an alignment typically requires tolerances in the submicrometer scale. Thus, the development of automated high-accuracy alignment methodology is highly desirable. In this context, the lack of (i) standardized photonics component designs and connections and (ii) standardized alignment and joining methodologies constitutes an additional difficulty, which may eventually be resolved by incorporating robotic manufacturing methods involving machine-assisted alignment and attachment processes that facilitate large-volume manufacturing.

To summarize, some of the important topics in photonics assembly and packaging (related to adhesive bonding) include precision placement, alignment, dimensional tolerances, component coupling, adhesives and their selection, thermal management, aspects of adhesive dispensing techniques, hermeticity issues, stresses in adhesive joints, reliability issues, monitoring the curing process, characterizing the adhesive joint, and automation concepts.

CHAPTER 2

Photonics Components, Assemblies, and Devices

CONTENTS

2.1. INTRODUCTION

In general, one can distinguish between two main types of optical and fiber-optic components, that is, *active* and *passive*. Active components generally relate to those requiring some type of power (and electronics), and their main function is directed at producing, controlling, and detecting the electromagnetic radiation. Active components include (i) semiconductor lasers and light-emitting diodes that produce light for various applications, (ii) photodetectors (or receivers), (iii) modulators, (iv) amplifiers, and (v) switches, and these typically require the integration of electronics and wiring into a package using traditional assembly methodologies. On the other hand, passive components typically do not have any electronics or power associated with them. Their main function is to direct, focus, filter, and divide/combine the light signals traveling, for example, through the optical fiber, and these typically include lenses and prisms, mirrors, couplers, isolators, and wavelength division multiplexers and demultiplexers.

Some examples of a wide range of applications of adhesives in assembling optical, fiber-optic, and optoelectronic structures and devices include lens bonding to various structures and devices, connector assembly, fiber splicing, fiber pigtailing, and fiber attachment to various structures and substrates (e.g., fiber-to-ferrule, fiber-to-glass, fiber-to-ceramic). These components will be described below (the terms of this description are also briefly outlined in the Glossary at the end of the book).

It is useful to briefly review some of the basic relevant concepts in optics and properties related to electromagnetic radiation. As mentioned in Chapter 1, optics is concerned with phenomena associated with electromagnetic radiation in the wavelength range between the vacuum ultraviolet (about 40 nm) and the far infrared (1 mm).

In the description of optical phenomena, it is often convenient to treat waves propagating through space as light rays, that is, lines normal to the waves and propagating in the direction of the advancing wave front. Thus, both the description of such rays and the optical design involving different components can be greatly simplified. One example is using a relatively simple equation related to bending (refraction) of light as it passes through a boundary between two media of differing refraction index (n), such as air and glass (see Figure 2.1). The law describing this behavior of light, referred to as Snell's law (expressed as $n_1 \sin \theta_1 = n_2 \sin \theta_2$), describes the change in the direction of a light ray at a boundary between two media of differing refraction index (n_1 and n_2). The angles of incidence and refraction (θ_1 and θ_2, respectively) are measured from

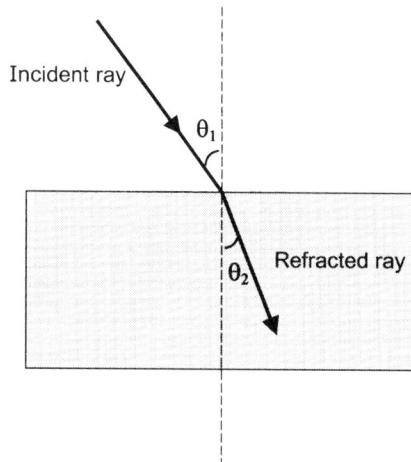

Figure 2.1 Schematic representation of refraction of light as it passes through a boundary between two media of differing refraction index (n), such as air and glass.

the normal to the surface. (Note that when a light ray is passing from low-to-high refractive index, it is bent toward that normal, whereas when the ray is passing from high-to-low refractive index, it is bent away from the normal.)

The refractive indices for selected materials are listed in Table 2.1.

It is important to note that, in general, the refractive index of light-cured polymers can be adjusted (between about 1.3 and 1.6) with great accuracy by blending different polymers, as well as by copolymerization (e.g., Eldada and Shacklette, 2000). Also note that manufacturers generally list the refractive index of the specifically designed adhesives that are referred to by their alphanumeric codes (which are different for different manufacturers).

The guiding of light in optical fibers is made possible by a phenomenon referred to as *total internal reflection* (TIR). The requirement for TIR is that the ray of light be incident on a dielectric interface from the high-refractive-index side to the low-refractive-index side. Figure 2.2 demonstrates various cases of incidence angles. For the case of propagation of a ray from a high-refractive-index medium (n_1) to a low-refractive-index medium (n_2) at an angle θ_1 that is smaller than the critical angle θ_c, a portion of light will be reflected back to a high-refractive-index medium and another part of light will be refracted into a low-refractive-index medium (see Figure 2.2a) according to Snell's law (i.e., $n_1 \sin \theta_1 = n_2 \sin \theta_2$). For the case of θ_1 reaching θ_c, $\theta_2 = 90°$, where the critical angle $\theta_c = \sin^{-1}(n_2/n_1)$, and the refracted ray will propagate along the boundary surface (see Figure 2.2b). For further increase in incidence angle to some value θ_3 that is greater than θ_c, the ray is totally reflected back into the higher refractive index medium, that is, satisfying the condition of total internal

Table 2.1
Refractive Indices of Selected Transparent Materials

Material	Average index of refraction
Silica glass	1.46
Borosilicate glass	1.47
Soda-lime glass	1.51
Quartz (SiO_2)	1.55
Dense optical flint glass	1.65
Lithium niobate ($LiNbO_3$)	2.2
Diamond	2.42
Polymers (general range)	1.3–1.6
Polypropylene	1.49
Polyethylene	1.51
Polystyrene	1.60

Note: The refractive index varies somewhat with the photon wavelength.

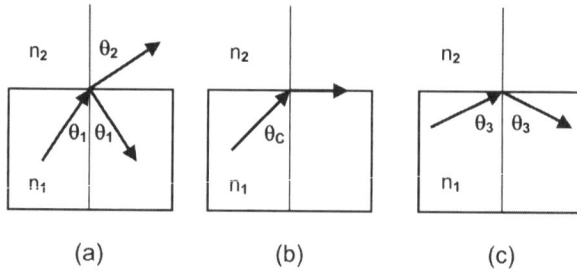

Figure 2.2 Refraction and reflection at the interface between two media with different indices of refraction ($n_1 > n_2$): (a) incident angle $\theta_1 < \theta_c$, (b) incident angle equals θ_c (i.e., the critical angle), (c) incident angle equals θ_3, which is greater than θ_c (corresponding to total internal reflection).

reflection (see Figure 2.2c). It should be noted, however, that light leakage between two closely spaced media could occur, leading to a possible evanescent wave coupling between fibers in a bundle.

2.2. OPTICAL COMPONENTS

As noted previously, the main function of passive components is to direct, focus, filter, and divide/combine the light signals traveling through the optical fiber and/or component that typically include lenses and prisms, mirrors, couplers, isolators, and wavelength division multiplexers and demultiplexers. These are typically mounted as individual components with great precision in various optical instruments and systems employing various means, including adhesive bonding. In this context, some of the important issues of concern are mounting-forces-induced stress buildup within optical components, resulting in modification of various material properties, and the effects of temperature variations on the bonded system that can cause stresses in improperly designed bonds.

In general, a lens (as an optical system) can be defined as a system for distributing and collecting light in a specific manner. A lens system, in principle, may incorporate several individual lenses having various sizes and shapes. In addition, a lens as a complete optical system may, in fact, incorporate a variety of individual lenses, mirrors, filters, prisms, diffraction gratings, and even moving (e.g., rotating) optical components. In addition, some optical systems may also incorporate a light source or a photodetector. These systems may operate in the infrared and ultraviolet wavelength ranges, in addition to the visible range; thus, materials considerations also have to be made to ensure the suitability of the optical components for the ranges other than the visible range (which is between about 400 and 700 nm).

Some of the applications of lens-based optical systems include microscopes, telescopes, binoculars, various camera lenses, medical instruments (e.g., endoscopes) used for surgery, and laser-based optical readers for compact discs (CDs).

There are many types of lenses having different sizes and shapes, and, as mentioned above, various combinations of lenses are employed in a wide range of optical systems in order to (a) bend the light in a desired manner or (b) form images. Two basic types of lenses are (i) a focusing (or converging, or positive) lens and (ii) a diverging (negative) lens (see Figure 2.3). These are also referred to as the *convex lens* (or biconvex lens) and the *concave lens* (or biconcave lens). Convex lenses are typically employed in combination with semiconductor light sources and detectors (e.g., for collecting and focusing light onto a miniature photodetector). The lenses shown in the figure are symmetric with similar (but opposite in sign) curvatures of the opposite sides (however, in many cases, individual lenses are not symmetric). It should be emphasized that the main characteristics that affect the route of light passing through the lens system include the properties of the lens glass material, its thickness, and curvatures.

Some of the issues of concern in lens design include spherical aberration, that is, the fact that, in geometrical terms, not all the rays from a zero-dimensional point object (imaged through a lens) focus to a single zero-dimensional image point. Spherical aberration is the result of the fact that, in the case of a simple

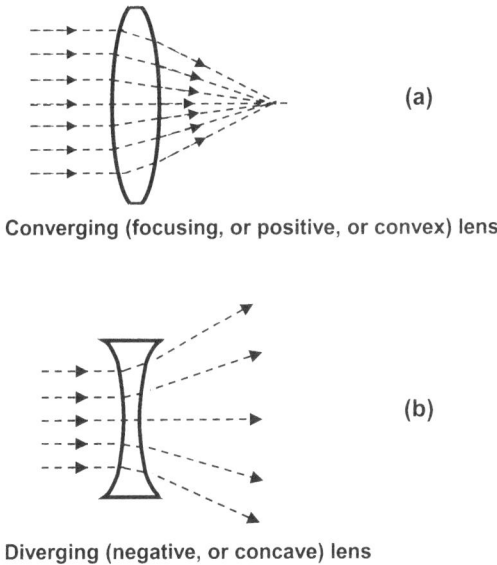

(a)

Converging (focusing, or positive, or convex) lens

(b)

Diverging (negative, or concave) lens

Figure 2.3 Converging (a) and diverging (b) lenses.

lens (or mirror) with a spherical surface, the rays, which arrive at different heights on it, do not bend in a similar manner, and thus these rays focus at somewhat different distance along the lens axis.

2.3. FIBER-OPTIC COMPONENTS

The main types of optical fibers, based on the material's type, are *glass optical fiber* and *plastic optical fiber*. Both glass optical fiber and plastic optical fiber consist of a cylindrical *core* surrounded by *cladding* (see Figure 2.4a). The core, the transmitting part of the fiber, has a higher index of refraction than the cladding, which constitutes the outer layers of the fiber. (Note that in the so-called plastic-clad silica fiber-optic cable, a plastic cladding surrounds a glass core.) It should be emphasized that in order to maintain the incident light rays reflecting at the boundary between the core and the cladding (and ensuring their propagation down the fiber), the cladding always has a smaller *refractive index* than the core. This is based on the phenomenon of *total internal reflection* (TIR);

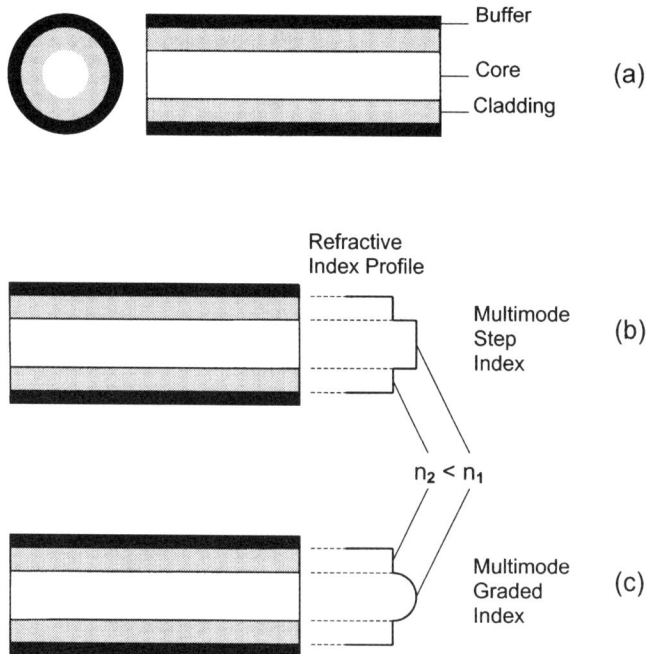

Figure 2.4a   Structures of (a) the basic fiber-optic cable and (b) and (c) the two basic types of optical fibers.

that is, no refraction can occur (thus, the light is totally reflected) if the angle of incidence of light trying to cross from a higher index medium to a lower index medium is greater than the critical angle (see Section 2.1). Thus, the light incident on the interface between a core and a cladding at an angle greater than the critical angle is reflected back into the core. However, as a result of destructive interference, only light with specific wavelengths, which are referred to as *modes*, can propagate through a given optical fiber. An outer polymer jacket (buffer) is used to protect the fiber from damage.

A fiber core is made of high-purity silica glass, and its diameter typically varies between about 5 (for single-mode fiber) and 100 (for multimode fiber) μm.

A higher refractive index of the core (as compared to the cladding) is realized by doping the core with a controlled amount of a dopant, which for pure silica glass (SiO_2) is germanium, having a greater total number of electrons than silicon. The greater total number of electrons in the core results in the reduced speed of light of the core glass and, thus, in the increased refractive index. As mentioned above, in the case of the refractive index of the core being greater than that of the cladding, light propagating in the core undergoes total internal reflection for light rays incident on the interface between the core and the cladding at an angle greater than the critical angle θ_c.

The basic types of optical fibers that are distinguished on the basis of the propagation modes are (i) *single-mode* optical fiber and (ii) *multiple-mode* optical fiber, and the latter is also distinguished as *multimode step-index* (the refractive index of the core is constant) and *multimode graded*-index (having variable refractive index) optical fibers (see Figure 2.4b). In a step-index optical fiber, the refractive indices of both the core and the cladding (note again that the refractive index of the core is always greater than that of the cladding) are uniform across their respective cross sections, and the light propagates along the optical fiber in a straight line. In a graded-index (GRIN) optical fiber, the refractive index of the core varies with the core radius in such a way that the refractive index gradually decreases away from the center of the core to the outer radius. Thus, the light, propagating along the optical fiber, is refracted back toward the axis of the fiber and has a curved trajectory (see Figure 2.4b).

Such an optical fiber can be used individually as a communication fiber (i.e., by transmitting pulsed optical signals), or it can be used in bundles for the transmission of light or images (see also, below, the description related to *coherent fiber bundle* and *incoherent fiber bundle*).

Employing such optical fibers, one form of communication is realized by rapidly switching the light "on" and "off," thus providing a means of sending short pulses (resulting in a digital signal) of light down the fiber. Depending on the core diameter and the wavelength of the light, the light pulse can propagate in several different pathways, referred to as *modes*. The data are transferred

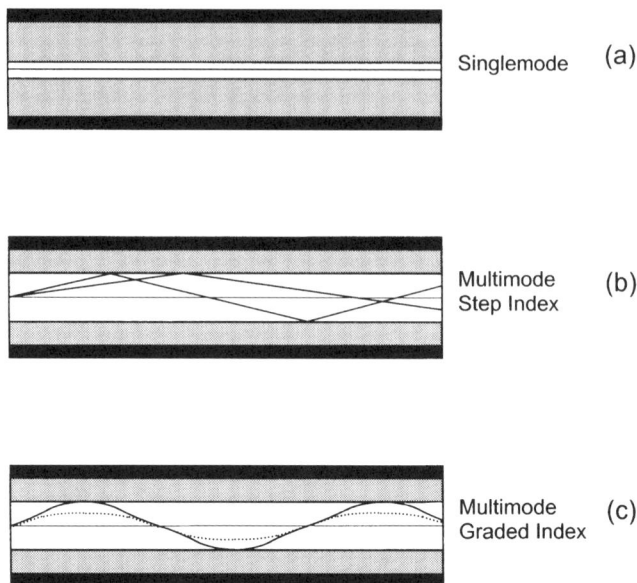

Figure 2.4b   Trajectories of light rays in (a) single-mode, (b) multimode step-index, and (c) multimode graded-index fibers.

through the optical fiber by switching the light "on" and "off" at various pulse rates. An important characteristic of an optical cable is the *bandwidth*, defined as the amount of data that can be carried by a cable (measured in bits per second), which is always much higher for a single-mode fiber than for a multimode fiber.

It is also important to distinguish between the *coherent fiber bundle* and the *incoherent fiber bundle*. In the former, a large number of single optical fibers are aligned and fused together in an ordered assembly, so that the fibers are ordered in precisely the identical manner at both ends of the bundle, thus making it suitable as an image guide. In the case of an incoherent fiber bundle, optical fibers are also assembled together, but they are not ordered in the identical manner at both ends, and thus they cannot transmit images, but can be employed as efficient light guides, having larger diameter individual fibers as compared to an image guide, which typically has lower diameter individual fibers for higher resolution.

One of the critical issues in fiber-optic communication is that of the intensity loss in optical fibers. The main causes of such a loss are *attenuation, connection losses*, and *bending losses*. *Absorption* and *scattering* contribute to attenuation losses in optical fibers, with the former being due to the absorption of light by molecules (such as OH$^-$ ions) in the fibers. The scattering is typically due to the small refractive index variations in the glass (referred to as Rayleigh scattering that follows the λ^{-4} dependence). These types of losses, which are dependent on wavelength, result in specific useful (low-absorption) *windows* suitable for fiber-optic communication purposes, for example, at 1.3 μm and 1.55 μm. The connection losses are typically due to (i) optical fiber connection misalignment, (ii) optical fiber mismatch, or (iii) reflections from the optical fiber ends. The bending loss, as the term indicates, is due to optical fiber bending that results in the refraction of the light rays incident (at the interface between the core and the cladding) at an lower angle than the critical angle.

An important characteristic of an optical system is the *numerical aperture*, which is related to the full *acceptance angle* 2θ of the cone of light rays that is capable of passing through the system (see Figure 2.5). Optical fibers are characterized by the *acceptance angle*. The numerical aperture of a fiber is NA = sin θ, and it is determined by the refractive indices of its core and cladding, that is, NA = $(n_1^2 - n_2^2)^{1/2}$. The rays entering a fiber at an angle equal to or less than the acceptance angle will be reflected internally and propagate down the fiber (note that, in this case, the light is propagating by the total internal reflection at the interface between the core and the cladding). However, the rays incident at angles greater than the acceptance angle will not be reflected internally (these will pass through or be absorbed by the cladding) and thus will not be guided in the fiber core. The numerical aperture essentially describes the ability of an optical fiber to collect light, and it defines the maximum angle to the axis of the fiber at which light is collected and propagated through the fiber. Similarly, the numerical aperture also represents the angular spread of light from a central axis, which is emitted from a light source or collected by a light receiver. Typical values of the numerical aperture are in the range between about 0.2 and 0.6.

Chromatic dispersion is another important phenomenon in optical communication systems. It relates to the fact that the propagation speed depends on wavelength. As mentioned above, the communication in such systems is realized by rapidly switching the light "on" and "off," thus providing a means of sending short pulses (resulting in a digital signal) of light down the optical fiber. However, if two pulses are sufficiently close to each other, they will ultimately

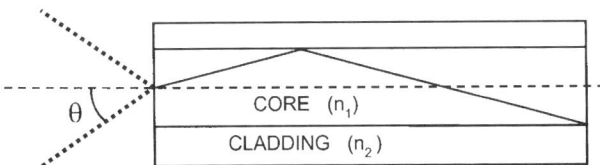

Figure 2.5 Acceptance angle of an optical fiber.

blend together; thus (for an optical receiver), it will become impossible to differentiate between them. Another view of chromatic dispersion relates to the fact that, in a typical light source, there are components at different wavelengths (i.e., spectral width), which essentially results (recall that the propagation speed depends on wavelength) in a pulse broadening and the associated broadening in time width of a propagating pulse. This phenomenon can be evaluated in a relatively simple manner by measuring the propagation time of light at different wavelengths through a fiber-optic cable.

The sizes of fiber-optic cable are typically given by two numbers, that is, the core size (given first), followed by the cladding size. For example, 100/140 indicates a core diameter of 100 μm and a cladding diameter of 140 μm. Although the larger core would ensure more light being coupled into it, an excessive amount of light may also result in saturation problems for the receiver (photodetector).

At this juncture, it is important to note the application differences between the glass and the plastic optical fibers. Glass optical fibers are relatively much more expensive with respect to their installation and maintenance. In addition, although very effective for underground installations, glass optical fiber is too delicate for horizontal subsystems. On the other hand, the strength of the plastic optical fiber is sufficient for its relatively easy installation and pulling through various systems and structures. Thus, plastic optical fibers can be employed in a wider variety of relatively inexpensive applications, but are presently not widely used (their spectral transmission properties are different from those of glass fibers).

In practical applications, optical fibers are typically joined together in order to (i) make longer lengths of fiber; (ii) repair broken fibers; and (iii) connect other devices such as filters, amplifiers, and so on. (In addition, in order to be able to link up with various components and devices, the ends of the fiber are routinely fixed with suitable connectors.) Optical fibers are typically joined employing two basic methods, that is, the permanent joint (often referred to as a *splice*) and the nonpermanent joint (referred to as a *connection*). One of the important efforts in joining two fibers is to minimize the loss of light at the joint.

Splicing is referred to a *permanent joining method* (without a connector) of the ends of similar optical fibers. Examples of the two main methods of splicing are thermal fusing and mechanical splicing. Thermal fusing is realized by the application of sufficient localized heat for melting or fusing the ends of the optical fiber cables, resulting in a single continuous optical fiber cable. In this process, the ends of the fibers are typically aligned by employing either manual or automatic methods. In the former, micromanipulators and a microscope are used, whereas in the automatic process one can either use cameras for monitoring or, alternatively, optimize light transmitted through the splice by adjusting the positions of the fibers. Following the alignment, the two fiber ends are melted jointly by employing, for example, an electric arc. The mechanical splicing method employs a device (typically made of glass) that facilitates the alignment of the two fibers. In this case, prior to positioning the fibers inside the splice, it is filled with an optical adhesive (usually an epoxy) having matching refractive index to the fiber core. The fibers are then adjusted (by optimizing the light transmission), secured in position, and finally exposed to UV light for curing the adhesive. It should be noted that, typically, fusion splicing provides a better joining method (transmission losses as low as about 0.02 dB, as compared to the loss of about 0.1 dB for the mechanical splice). Some of the mechanical means include using grooves (using V-block) for precise positioning of the fibers, establishing contact with each other, followed by the application of epoxy for permanent positioning. One can also employ a precision sleeve, where the two fibers are inserted at opposite ends of the sleeve and aligned, followed by an injection of an index-matching epoxy into the sleeve to facilitate the permanent splice (see Figure 2.6).

In these processes, in order to prevent any transmission losses, the ends of the fiber must be accurately lined up with each other. Typical alignment errors during the splicing of optical fibers include (i) lateral misalignment, (ii) angular misalignment, and (iii) axial misalignment (see Figure 2.7); in addition, such fiber characteristics as the differing fiber core sizes and the quality of fiber end finish can also contribute to transmission losses.

In addition to *splice loss*, there are other types of losses, which are outlined below. These include *connector loss, coupler loss*, and *light-source-to-fiber coupling loss* and *fiber-to-receiver coupling loss*. (Note that the fiber-to-receiver coupling loss is typically negligible, since the area and the numerical aperture of the detector are typically greater as compared to the fiber.) In addition, as described above, the intensity loss in optical fibers also occurs due to *absorption* and *scattering*, which contribute to attenuation losses in optical fibers.

Connectors do not provide a permanent joint; these can be detached and used again. These include the so-called *butt connector* and *lens connector*. In the former case, each fiber end is positioned in the ferrule, and the corresponding two ferrules are fixed inside a sleeve that facilitates the alignment of ferrules; such connectors can also be configured as a *multichannel connector* for connecting

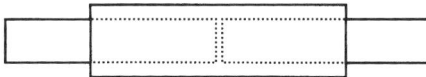

Precision sleeve

Figure 2.6   Schematic diagram of a precision sleeve.

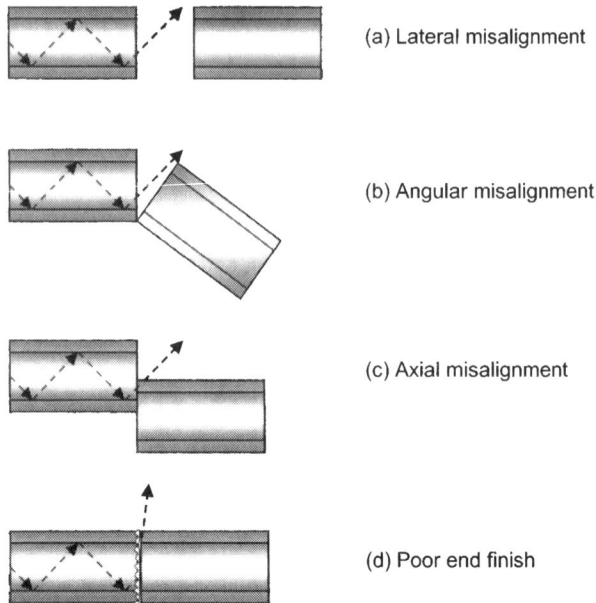

(a) Lateral misalignment

(b) Angular misalignment

(c) Axial misalignment

(d) Poor end finish

Figure 2.7   Alignment errors and fiber characteristics that affect transmission losses.

several fibers. In a lens connector, a lens configuration collimates the beam of light from one fiber into another.

Important components of optical networks, which also require specific coupling and switching configurations, include *optical couplers, Y-junctions,* and *switches.*

Optical couplers (e.g., directional coupler and star coupler) are designs for splitting the light. This includes, for example, the four-port directional coupler (2×2 coupler), in which the light received at a given port is split between two other ports (typically 50%–50% split). Such couplers are prepared by fusing and tapering two fibers together.

Y-junctions (see Figure 2.8) (e.g., 1×2, 1×4, 1×8 couplers) constitute a principal component in optical networking. Preferably, the incoming light is split equally between the two arms of the junction.

Star couplers (see Figure 2.9) are optical couplers having more than four ports. For example, in a transmission star coupler, shown in Figure 2.9, the light arriving at port 1 is split equally through ports 7 through 12.

Switches, which send the optical signals in a specific direction, are important components of optical networks. The two basic types of switches are a *two-position switch* and a *bypass switch.* In a two-position switch, the optical signal

(a) Y-junction or 1×2 coupler

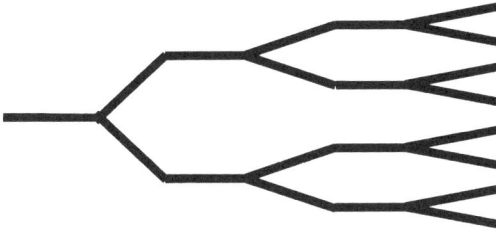

(b) Y-junctions or 1×4 or 1×8 coupler

Figure 2.8   Schematic diagram of types of Y-junctions: (a) 1×2 coupler, which ideally facilitates the incoming light to split equally between the arms; and (b) 1×4 or 1×8 coupler.

arriving at port A, for example, can be switched to either port B or port C. Typically, the two-position switch employs three optical fibers in conjunction with associated collimating lenses and a prism.

An additional critical component in fiber-optic assemblies is the *source coupling*, that is, coupling of light from a light source into an optical fiber (preferably with minimal loss). An efficient coupling of the diverging optical beam with an optical fiber can be realized by employing appropriate lenses (see Figure 2.10).

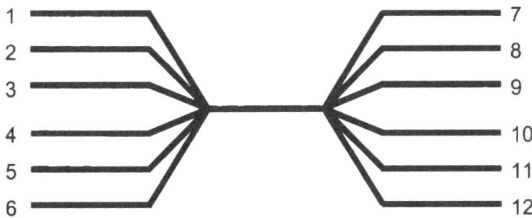

Figure 2.9   Transmission star coupler.

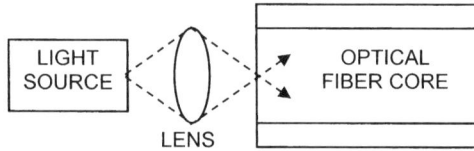

Figure 2.10 Efficient coupling of the diverging optical beam with an optical fiber by employing a lens.

2.4. OPTOELECTRONIC INSTRUMENTATION

Some of the essential components in photonics communications technology are *light-emitting diodes* (LEDs), *diode lasers*, and *photodetectors*, which can be made sufficiently small and efficient for various applications. Such devices are typically based on various semiconductors. The main utility of semiconductors in such applications is based on the effects produced by the introduction of various dopants that facilitate the formation of built-in electric fields and corresponding junction devices that can be used as light-detecting and -emitting devices. A light-emitting diode is essentially a semiconductor device, which emits light from the p–n junction under the applied forward bias, whereas a laserl is a light source generating coherent and near-monochromatic light. A photodetector is a semiconductor device (an optoelectronic transducer), which produces a photocurrent in response to absorbed incident optical power; thus, it can be used as a detector in a fiber-optic cable data link.

A light-emitting diode (LED) is a semiconductor p–n junction device, which employs radiative recombination and subsequent emission of light from the forward-biased junction; in this process, electrons injected in the conduction band fall into the valence band and release the extra energy that is emitted as a photon; in such a process, photons are emitted in random directions, hence it is referred to as incoherent light.

The same process is used in a semiconductor *diode laser*, which differs from an LED in (i) the operating current (i.e., much greater current for achieving *optical gain*) and (ii) having opposite ends cleaved parallel to each other; this results in the formation of aligned mirrors that reflect the generated light back and forth for achieving amplification of light and generation of *stimulated emission* [in this process, conduction-band electrons, which are stimulated by photons, fall into the valence band, and photons emitted as the result of this stimulated emission constitute coherent light, i.e., (i) the emitted photons have the same energy as the incident photons and (ii) they are in step with each other]. Thus, as compared to LEDs, semiconductor diode lasers typically provide light output with narrower wavelength range, higher power, and more directionality.

Various semiconductors that are generally employed in optoelectronic applications include the III–V compounds, such as GaAs, InP, GaP, InAs, and GaN, and their corresponding ternary and quaternary alloys play an especially important role in such applications. This is basically due to the wavelength range at which these materials emit and absorb light efficiently (this is determined by the magnitude of the semiconductor *energy band gap*, i.e., the energy separation between the valence and conduction bands). For example, $Al_xGa_{1-x}As$ is suitable for light emission in the range between about 750 and 900 nm; $GaAs_{1-x}P_x$, emitting at about 650 nm, is suitable for plastic fibers; and $Ga_xIn_{1-x}As_yP_{1-y}$ is suitable for light emission in the range between about 1300 and 1700 nm (i.e., within the most important range of fiber-optic communication).

The efficiency of an LED depends substantially on self-absorption of the generated light in the material (i.e., it depends on the absorption coefficient of the material at the emission wavelength). The absorption of light can be minimized by employing a heterojunction device with the top layer having a wider energy gap. One example of such a structure is based on an $Al_xGa_{1-x}As/GaAs/Al_xGa_{1-x}As$ system that facilitates the formation of a double-heterostructure (DH) LED, in which the top $Al_xGa_{1-x}As$ layer acts as a window allowing the generated light to be emitted with no self-absorption. In fact, most practical optoelectronic devices incorporate various types of heterojunctions for purposes such as providing the transparency to light generated in the active region (as mentioned above).

In general, light-emitting diodes and diode lasers have great advantages (as compared to other sources of optical radiation), since they (i) provide radiation with higher efficiency, (ii) have longer operational lifetimes, and (iii) emit radiation in a narrow wavelength range.

In some applications (e.g., for efficient delivery of optical radiation for processing, such as curing of various targets), in order to produce high-power light-emitting device assemblies, a large number of such light-emitting devices are configured next to each other in fixed coordination, such as, for example, linear configuration. In such a case, it is essential that the light emitted by the said array be delivered to a desired location with the desired intensity and irradiance distribution.

In general, besides the emission wavelength range, LEDs are characterized by the direction of emission as well. Thus, depending on device geometry (as well as internal structure), LED types are distinguished between the *surface-emitting* and the *edge-emitting* types (see Figure 2.11). The surface-emitting LEDs are employed in common light source applications. For optical communication, the emitted light has to be coupled efficiently to a fiber or wave guide with minimum losses; this requires the emission of light only in the direction of the cable. The requirement for high intensities demands confinement of the

Light output

↑

(a) Surface - emitting LED

```
                                    +
┌─────────────────────┐
│       p-type         │
├─────────────────────┤  — Junction          Contacts
│       n-type         │
└─────────────────────┘
                                    −
```

Junction ─────

Contacts

(b) Edge - emitting LED

```
                        +
┌─────────────────────┐
│       p-type         │
├─────────────────────┤  ⇒  Light output
│       n-type         │
└─────────────────────┘
                        −
```

Active region ─────

Light output

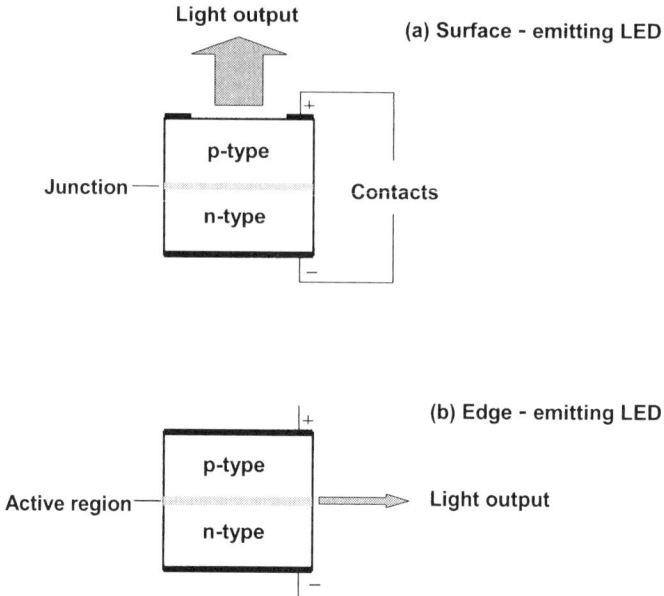

Figure 2.11 Schematic diagram of (a) surface-emitting LED and (b) edge-emitting LED. Note that light is actually emitted in different directions, but typically the manner of device packaging allows light output in one direction only.

light emission to a small region with high injection ratios. This is typically real-ized by employing edge-emitting LEDs. With double-heterostructure devices, it is possible to realize very high injection ratios by employing injection from a wider energy gap semiconductor to a narrower energy gap material, where the injected carriers become trapped. In addition, such a heterostructure confines the light from both sides due to the change in composition (and, correspondingly, in refractive index) along the junction. (Note that the refractive index of GaAs is greater than that of AlGaAs, thus providing the condition for confinement of light in the GaAs active region due to total internal reflection.) Similar structures can be realized by employing $Ga_xIn_{1-x}As_yP_{1-y}$/InP heterojunctions.

The *photodetectors* that are employed in fiber-optic communication are also based on semiconductor junction devices (including the p–i–n or PIN photodi-odes and the avalanche photodiodes). In this case, unlike light-emitting devices that are forward-biased, in order to make these photodiodes more sensitive and faster, semiconductor junctions in photodetectors are reverse-biased (reverse bias implies that a positive bias is applied on the n-side of the diode, whereas a negative bias is applied on its p-side). The basic principle of a photodetector is related to the absorption of incident photons that results in the generation of electron–hole pairs with (i) subsequent separation of these charges through the junction in opposite directions and (ii) the flow of current in the device. Since the absorption process of photons with a given energy by a semiconductor depends on its energy gap, different semiconductors are suitable for detecting light having different wavelengths. Some of the important semiconductors for photodetectors employed in fiber-optic systems (and the corresponding approximate wavelength ranges of detectors) include Ge (500–1550 nm), Si (400–1100 nm), GaAs (400–860 nm), and $Ga_xIn_{1-x}As$ (900–1700 nm). It should be emphasized that each of these photodetectors has a specific dependence of the photoresponse on wavelength. Thus, for example, the Si detector has a maximum photoresponse at about 900 nm, whereas $Ga_xIn_{1-x}As$ has a nearly flat responsivity between about 1200 and 1700 nm and is typically optimized for such important wavelengths (for optical fiber communication) as 1300 and 1550 nm.

2.5. OPTOMECHANICAL SYSTEMS

As mentioned in Chapter 1, the application of optical adhesive bonding in various *optomechanical* systems requires precision and secure mounting of optical components. It should be emphasized that, in this case (i.e., optomechanical adhesive bonding), more stringent requirements are needed than in other adhesive bonding applications, since it is critical that (in order to prevent any optical distortion) adhesively bonded components remain precisely positioned during their operating lifetime.

Some typical materials employed in optomechanical components include aluminum, stainless steel, and brass, and their choice in specific applications is often determined by a range of properties, such as:

(i) The *thermal expansion coefficient* (note that the variations in temperature result in changes in the size/shape of a component, and these changes typically depend on the component size, the temperature variation range, and the specific material used)

(ii) The *stiffness* (which is proportional to the modulus of elasticity)

(iii) The *specific stiffness* (i.e., the stiffness per density of the material), which has to be taken into consideration in relation to the vibration resistance of the system (specifically, greater specific stiffness leads to better vibration isolation)

In many cases, a certain trade-off in materials characteristics is considered. For example, although the thermal expansion coefficient of aluminum is greater by about a factor of 2 as compared to stainless steel (which makes stainless steel more advantageous in some applications), the thermal conductivity of aluminum is also substantially greater (as compared to stainless steel), which is highly advantageous for mounting such power sources as diode lasers and dissipating heat from them more efficiently and thus reducing possible distortions associated with thermal gradients along the component. (Note that such a distortion is proportional to the ratio of the coefficient of thermal expansion and the coefficient of thermal conductivity, and that ratio is about three times smaller in the case of aluminum as compared to stainless steel.)

2.6. PHOTONICS ASSEMBLY AND PACKAGING

As an illustration, an example of photonics assembly and packaging is depicted in Figure 2.12, which presents the case of an optical component mount, in which various components can be mounted to a base surface such as inside a butterfly package in a laser diode. The optical components are typically attached with suitable adhesives. Such optoelectronic devices are commonly packaged in hermetic butterfly housing made of low-thermal-expansion materials such as Kovar or a ceramic. Due to the low thermal conductivity of Kovar, heat sinks are typically brazed to the base of the package, which may also contain the photodiode monitor, thermoelectric cooler, thermistor, ball lens, and optical isolator. Some of the important requirements in such a case include stable optical coupling, effective thermal pathway, no outgassing, passive alignment (preferable), and hermetic enclosure (the lid is typically glass or ceramic).

As noted in the Preface, in the context of the present description, assembly is related to processes such as alignment, adjustment, and physical attachment of materials and components associated with forming the end product, whereas packaging relates to a process (e.g., hermetic sealing) for providing a means of protection of the components, assemblies, and devices (as end products) from the environment. The requirements for assembly and packaging are different; thus, different types of adhesives may be required for each of them.

An important characteristic in packaging is the thermal expansion coefficient (see Table 2.2).

2.7. SUMMARY

The two main types of optical and fiber-optic components are *active* and *passive*. Active components include (i) semiconductor lasers and light-emitting diodes, (ii) photodetectors, (iii) modulators, (iv) amplifiers, and (v) switches, and the main function of these components is related to producing, controlling, and detecting electromagnetic radiation. Passive components typically include lenses and prisms, mirrors, couplers, isolators, and wavelength division multi-

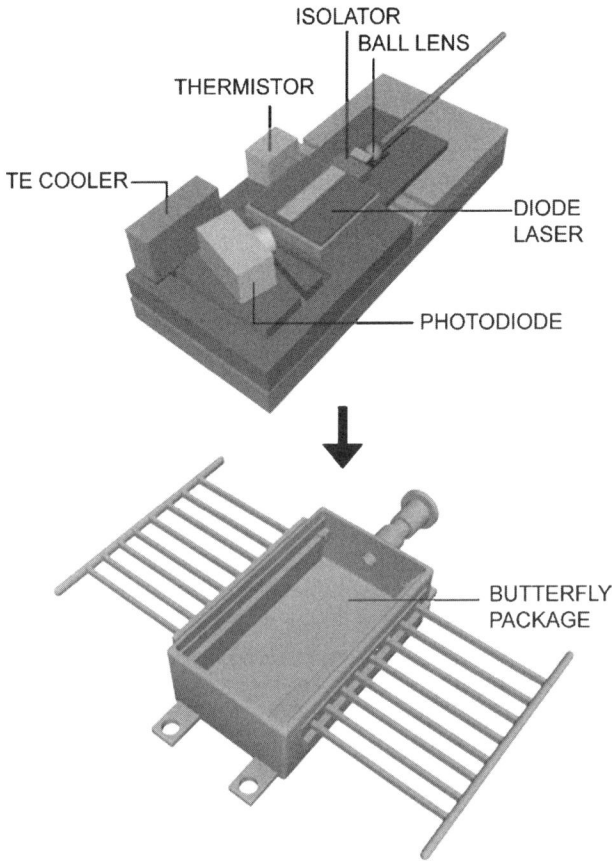

Figure 2.12 Schematic illustration of an example of photonics assembly and packaging presenting the case of a diode laser. In this type of application, the various components are placed on a silicon substrate and bonded in place with suitable adhesives (note that alignment is crucial in this application). Such optoelectronic devices are commonly packaged in hermetic butterfly housing made of low-thermal-expansion materials such as Kovar or a ceramic. Due to the low thermal conductivity of Kovar, heat sinks are typically brazed to the base of the package, which may also contain the photodiode monitor, thermoelectric (TE) cooler, thermistor, ball lens, and optical isolator.

plexers and demultiplexers, and their main function is to direct, focus, filter, and divide/combine the light signals traveling, for example, through the optical fiber.

Some examples of applications of adhesives in assembling optical, fiber-optic, and optoelectronic structures and devices include lens bonding to various structures and devices, connector assembly, fiber splicing, fiber pigtailing, and fiber attachment to various structures and substrates.

Table 2.2

Thermal Expansion Coefficients of Selected Materials

Material	Thermal expansion coefficient ($\times 10^{-6}/^\circ$C)
Al_2O_3 (ceramic)	6.5
SiC (ceramic)	3.7
Silicon (single crystal)	4.0 (3)
Glass (soda lime)	2.8
Copper	16.8
Aluminum	22.4
Tungsten	4.5
Kovar	4.7
Sapphire	7.6
Gold	14.0
Epoxy	45–100
Urethane	60–250
Silicone	200–1000

The guiding of light in optical fibers is made possible by a phenomenon referred to as *total internal reflection* (TIR). The requirement for TIR is that the ray of light be incident on a dielectric interface from the high-refractive-index side to the low-refractive-index side.

The essential components in photonics communications technology include *light-emitting diodes* (LEDs), *diode lasers*, and *photodetectors*. These devices are typically based on semiconductors. A light-emitting diode is essentially a semiconductor device that emits light from the p–n junction under the applied forward bias, whereas a laser is a light source generating coherent and near-monochromatic light. A photodetector is a semiconductor device (an optoelectronic transducer) that produces a photocurrent in response to absorbed incident optical power; thus, it can be used as a detector in a fiber-optic cable data link.

CHAPTER 3

Fundamentals of Adhesive Bonding

CONTENTS

3.1. INTRODUCTION

In general, adhesive bonding is extensively employed in a wide variety of applications, such as bonded structures, electronics assembly and packaging, photonics assembly and packaging, medical device manufacturing, composite materials, automotive technology, and paints. There is an extensive literature on the subject, and some of the most recent books and reviews are listed in the Bibliography. This book is concerned with the adhesives science and technology related to photonics applications (i.e., photonics assembly and packaging), with the main emphasis on fiber-optic technology, which constitutes one of the pivotal technologies of the current advances in the telecommunications industry.

An adhesive is defined as a material capable of holding two other materials together in a practical manner by surface attachment that resists separation. This book is concerned with synthetic organic adhesives, which are typically composed of *polymers*, that is, *macromolecules* formed by the linking of many simpler molecules known as *monomers*. Typically, polymers are flexible, and they are able to spread and interact on the surface of the substrate material (this is highly important in practical applications of adhesives). The basic physical properties of polymers (e.g., melting point, viscosity, solubility, and tensile strength) are determined mainly by the strength of the intermolecular forces, the molecular weight, the polymer structure, and the flexibility of the polymer molecule.

In the context of this book, the formation of the polymer, that is, *polymerization* (defined as a chemical reaction resulting in the linking of molecules of monomers and forming of polymers through chain growth), can be achieved by a photoinduced reaction, leading to the conversion of the liquid monomer into a solid polymer (see, e.g., Fouassier, 1995). The process of the conversion of an adhesive from a liquid to a solid state is termed *curing*, and it can be accomplished, for example, by radiation curing (e.g., optical beam, electron beam, X-rays, microwaves). Optical radiation curing employs ultraviolet (UV), or visible, and or infrared (IR) photons. To summarize briefly, the main advantages of polymers include their flexibility, their capability to spread and interact on a material's surface, and their strength, which are all essential for the formation of an adequate adhesive bond.

It should be noted that most adhesives in practice are a combination of several constituents performing various important functions. These constituents may include:

(a) Adhesive base or binder (i.e., the active bonding agent)
(b) Solvents
(c) Initiator (for curing)
(d) Accelerators, inhibitors, or retardants (to control the curing rate)
(e) Fillers (to control properties)
(f) Carriers (for reinforcement)

Joining components and materials with adhesives has several advantages as compared to mechanical methods. These advantages include:

(a) Adhesives generally provide a continuous bond, and they distribute load more uniformly and over wider areas.
(b) In addition to bonding, adhesives can also seal against moisture penetration, and thus prevent corrosion.
(c) Adhesives can bond irregularly shaped surfaces more easily.
(d) Adhesives can make a bond more resilient to stress.

Adhesives are also advantageous in both cost and convenience as compared to mechanical means, and they are especially better suited to the assembly of fiber-optic components and structures if used properly (especially photocured adhesives).

The main disadvantages of adhesive bonding include:

(a) A more limited (as compared to mechanical methods) service temperature range, which is primarily related to the glass transition temperature and/or chemical degradation of the adhesives, resulting in loss of strength or adhesion

(b) The fact that adhesive strength may depend strongly on the condition of the surface of the adherend (i.e., the material being bonded)

(c) The fact that adhesive bonding may deteriorate in the presence of water and/or water vapor

It would be informative to compare the adhesive bonding method with other joining techniques, which are also typically employed in photonics applications. These include the two most important techniques, soldering and laser welding. They all have their specific advantages and disadvantages, which are outlined in Table 3.1.

Note that *laser beam welding* is a process that joins different parts by employing the heat generated by a laser beam directed onto the weld joint, whereas *soldering* is a process by which two (metal) surfaces are bonded together by means of an intermediary alloy (typically used for making electrical connections). The latter is also referred to as *soft soldering*, which should be distinguished from *hard soldering*, which is also referred to as *brazing* (note that the joining temperature for soft soldering is about 250 °C and that for hard soldering is about 1000 °C).

In general, the major topics of interest related to adhesion and adhesives science are:

(a) Theories and mechanisms of the adhesive bond

(b) Physical and chemical properties and characteristics of polymers relevant to adhesive bonding

(c) Adhesive intermolecular forces and surface science

(d) Determination of surface free energy of polymers

(e) Relevance of wettability and surface energetics in adhesion

(f) Surface preparation of materials for adhesive bonding

(g) Various methods of adhesion enhancement

(h) Molecular dynamics modeling of adhesion

(i) Curing of adhesives

(j) Characterization methods of adhesive bond and related interfaces

(k) Design and mechanical testing of adhesive joints

It should be emphasized that in practical cases many variables have to be considered in order to understand the specific adhesive bond, and no single theory, model, or simple relationships containing constants and materials parameters could explain all types of adhesive interaction and bonding, although it is certain that chemical interfacial interactions have a major contribution to the

Table 3.1

Comparison of Joining Techniques

Technique	Advantages	Disadvantages
Adhesive bonding Joining temperature: Room temperature–200 °C	Ability to join dissimilar materials Refractive-index matching in photonics Relatively low processing temperature (advantageous for heat-sensitive components) Possibility of re-work or alignment during cure Relatively inexpensive	Susceptible to moisture and outgassing Inability for component adjustment following bonding Limited service temperature range (related to the glass transition temperature and/or chemical degradation of adhesives) Nonhermetic
Soldering Joining temperature: ~250 °C	Applicable to a wide variety of metals Relatively easy rework (using solder re-flow) Hermetic	High (and prolonged) heat loads that may affect other components High-temperature strain Difficult to control shrinkage Need for metallization of optical fibers
Laser welding Joining temperature: ~1000–2000 °C	Strong joints Diversity of joints Relatively easy to automate Higher positioning speeds Controllability and reproducibility of the process Hermetic	Dissimilar materials (e.g., metal to glass) cannot be bonded Welded lines brittle at high temperatures The need for high weld energies and for precise control of energy delivered High cost of initial ownership Need for metallization of optical fibers

strength of various types of adhesive bonds. The major factors that influence adhesion include:

(a) Various interactions at different length scales (including the nanoscale)
(b) Surface morphology (note that real surfaces are not perfectly smooth)
(c) Deformation processes at different length scales

To explain the origins of adhesion, it is necessary to understand such processes at the atomic level and to evaluate the range of validity of continuum theories (Landman et al., 1990). This requires understanding of the atomic processes taking place at the interface between two materials that are brought together (or moved with respect to one another). In this context, the details on the relative importance and differences of the various scale levels from the macroscale to the microscale and to the nanoscale, as well as the transition between these levels, are crucial in interpreting various observations. It should be emphasized that with such scaling not only the measured materials properties at different scale levels may vary substantially, but the mechanisms underlying their physical properties and their interpretation can be different as well, related to a transition from bulk properties to surface and interface properties (Landman et al., 1990). Some specific issues of importance are those related to the atomic and molecular processes, which take place at the interface between materials and which have significant influence on adhesion, friction, lubrication, and nanoindentation.

The fundamental factors that determine the performance of adhesive bonding are:

(a) The physical and chemical properties of the *adhesive*
(b) The type of *adherend*, that is, the material being bonded
(c) The nature of the *surface* treatment for adhesion improvement
(d) The wettability of the surface
(e) The details of joint design

In practice, these factors (i.e., those related to adhesive, adherend, and surface) determine the service life of the bonded structure. An additional important factor, which determines the formation of a suitable adhesive bond, is the tendency of the adhesive to wet and spread on the surface of the adherend being bonded. The establishment of such a wetting results in the generation (across the interface) of adhesive forces that may arise due to several possible mechanisms (see the following sections for a detailed discussion).

The principal issues are related to control of the forces that govern adhesion and to identification of the key parameters and principles that are dominant in controlling the interface properties. In this context, some of the most important aspects of adhesive bonding include:

(a) The suitability of using adhesive bonding as a proper joining method for a given case
(b) The selection of the most suitable adhesive type for both the materials involved and the anticipated operating conditions
(c) Design of the adhesive joint
(d) Identification of the required surface preparation techniques for adhesion improvement

3.2. INTERATOMIC AND INTERMOLECULAR FORCES AT THE INTERFACE

In principle, the types of molecular and/or atomic bonds involved in the adhesion include chemical bonds, such as covalent, ionic, and metallic, as well as van der Waals and acid–base forces, including hydrogen bonding. Ionic bonding, which occurs as a result of electron transfer from one atom to another, is due to the Coulombic attraction between oppositely charged ions. In covalent bonding, the electrons are shared between neighboring atoms. In many cases, the electrons are not shared equally between the atoms, leading to partial ionization and mixed ionic and covalent bonding. In metallic bonding, the valence electrons, being free to migrate, are shared by all of the ions in the solid. The bonding in this case can be considered as an electrostatic interaction between the positive array of ions and the negative electron gas. Molecular or van der Waals bonding is due to the attractive force between two molecules as the result of electric dipole interactions; this bonding is relatively weak. The hydrogen bond is formed in a compound of hydrogen and strongly electronegative atoms, such as oxygen; in this case, the hydrogen is positively charged and can be shared between neighboring atoms or molecules. Thus, basically, the forces responsible for bonding are Coulombic in nature. (For a detailed discussion related to the types of bonds involved in the adhesion as a result of the presence of interatomic and intermolecular forces at the interface, see Israelachvili, 1991).

In the literature, the above forces are also referred to as (i) *primary* (or strong) bonds (i.e., ionic, covalent, and metallic bonds) and (ii) *secondary* (or weak) bonds (i.e., hydrogen and van der Waals bonds). It should also be noted that the types of bonds that are typically associated with polymers are covalent and van der Waals bonds.

Thus, in order to elucidate adhesive bonding, it is essential to consider the various possible interatomic and intermolecular forces and their importance in the understanding of the properties of surfaces and interfaces.

In general, the underlying origin of the surface forces is related to the Coulomb force occurring between electrically charged atoms or molecules. The potential energy between two electrical charges q_1 and q_2 separated by a distance r is expressed as

$$U = \frac{q_1 q_2}{4\pi\varepsilon r} \tag{3.1}$$

The force in this case is dU/dr, and it is expressed as

$$F = \frac{q_1 q_2}{4\pi\varepsilon r^2} \tag{3.2}$$

An important set of forces that are applicable to adhesive bonding is related to van der Waals forces, that is, dipole–dipole interactions (sometimes referred to as $1/r^6$ interactions), which may include:

(a) Permanent dipole–permanent dipole force
(b) Permanent dipole–induced dipole force
(c) Induced dipole–induced dipole force

The (permanent) dipole–(permanent) dipole force (also referred to as the *polar* or *Keesom force*) is due to the interaction between the permanent dipoles of two molecules. Such interactions depend on the orientation of the molecules, with the population of such orientations being dependent on the temperature. The potential energy of the interaction of two free-rotating dipoles, that is, the so-called *Keesom potential* is

$$U_K = -\frac{2\mu_1^2\mu_2^2}{3kT(4\pi\varepsilon)^2\, r^6} \tag{3.3}$$

where μ_1 and μ_2 are dipole moments ($\mu = ql$, where q is the magnitude of a virtual charge on the ends of the molecule and l is the distance separating the charges).

If a molecule with a permanent dipole (which is freely rotating) interacts with a polarizable molecule, the potential energy (referred to as the *Debye potential* or *induction potential*) is

$$U_D = -\frac{\mu^2\alpha}{(4\pi\varepsilon)^2 r^6} \tag{3.4}$$

where α is the polarizability of the nonpolar molecule and μ is the dipole moment.

The instantaneous transient dipoles (in nonpolar molecules), caused by the fluctuating distribution of electrons, produce attractive forces; in this case, the potential energy (which is also referred to as the *London dispersion energy*) of the interacting molecules depends on their polarizability, and it can be expressed as

$$U_L = -\frac{3}{2}\frac{\alpha_1\alpha_2}{(4\pi\varepsilon)^2 r^6}\frac{I_1 I_2}{(I_1 + I_2)} \tag{3.5}$$

where α_1 and α_2 are the polarizabilities of the molecules and I_1 and I_2 are their ionization energies.

The above three interactions are often grouped as van der Waals forces and, in general, they play a major role in various processes, including adhesion.

The different types of bonds discussed above have different bonding energies, which are summarized in Table 3.2.

Table 3.2

Summary of Types of Bonds and Their Approximate Magnitudes

	Type of bond	Bond energy (kJ/mol)	Comments
Primary (strong)	Ionic	600–1500	Bonding in crystals
	Covalent	100–1000	Bonding in crystals and cross-linked polymers
	Metallic	70–800	Welded joints
Secondary (weak)	Hydrogen	10–50	Sharing of positively charged hydrogen between neighboring atoms or molecules having lone pairs of electrons
	van der Waals		
	Polar	4–25	Dipole–dipole
	Induction	2–10	Dipole–induced dipole
	Dispersion	0.1–40	Fluctuating dipoles

3.3. SURFACE SCIENCE AND ADHESION

Surface and interface (defined as a boundary between different phases such as solid, liquid, or gas) properties have a major influence on adhesion. The interface properties, such as interfacial tension and interfacial adsorption, play a dominant role in adhesion science. *Interfacial tension* (related to the resistance to expansion of the interfacial area) describes the tendency of the interfaces to contract as the result of the system's drive toward minimizing the Gibbs free energy. *Interfacial adsorption* (related to the buildup of a foreign substance in the interfacial region) typically lowers the interfacial tension.

In general, adhesion is characterized by the interaction energy expressed in terms of surface free energies:

$$W_A = A_c \, \Delta\gamma \tag{3.6}$$

where A_c is the contact area (note that this area is not accurately defined in realistic cases because of surface roughness) and $\Delta\gamma$ corresponds to the gain, related to the surface free energy, due to the formation of the contact (and an interface) between two surfaces having surface free energies γ_1 and γ_2 and forming an interface with a surface free energy γ_{12}. One can also refer to $\Delta\gamma$ as the energy per unit area required for pulling apart the interface, and this is expressed as

$$\Delta\gamma = \gamma_1 + \gamma_2 - \gamma_{12} \tag{3.7}$$

One of the major topics in adhesion science is how to interrelate these quantities with such materials properties and external parameters as the atomic structure, chemical composition, surface contamination and treatment, temperature, and applied pressure. (It should be noted that in practical cases of adhesion the surfaces of materials are not perfectly smooth. This essentially indicates the difficulties involved with determining the contact area and an important concept of interface free energy in such cases.)

The adhesion is often expressed as $W_A = \gamma_1 + \gamma_2 - \gamma_{12}$, and it is referred to as the *thermodynamic adhesion*, describing the change in free energy as the result of the formation/partition of the interface. This may be distinguished from *fundamental adhesion* (i.e., the sum of all interfacial molecular interactions, on a molecular scale, between the contacting surfaces) and *practical adhesion* (related to the work required to separate a coating from the substrate; this depends on several factors, including fundamental adhesion, elastic properties of the materials, and intrinsic stress in the coating) (see, e.g., Mittal, 1976; Packham, 1996; Pignataro, 1998).

To summarize, adhesion may involve several factors, such as:

(a) Various interactions at different length scales
(b) Deformation mechanics at different length scales
(c) Surface morphology (see, e.g., Dürig and Stalder, 1998)

In this context, one of the most crucial steps is the formation of interfacial bonding, which follows the sequence of:

(a) *Wetting*, that is, a molecular contact between the adhesive and the adherend is established at the interface by a flow process

(b) *Adsorption*, that is, the formation of the adhesive bonding across the interface

(c) *Interdiffusion*, that is, the formation of physical cross-links (this occurs in the case of both the adhesive and the adherent being polymers with similar solubility parameters)

Wetting is defined as the process of spreading of a liquid onto a solid surface. This process is mainly controlled by the surface energy of the liquid–solid interface in opposition to the solid–vapor and liquid–vapor interfaces. Note that, in order to wet a surface of the adherend by the adhesive, the surface tension of the adhesive must be lower than that of the adherend. In general, it is important to consider both equilibrium and dynamic processes related to wetting, as well as the spreading of liquids on substrates.

As mentioned above, the microscopic features of the surface of a substrate are typically nonuniform. It is, therefore, essential that the adhesive completely wet the bonding area. There are two major factors that determine the extent that the surface can be wetted by the adhesive; these are (i) the viscosity of the material (adhesive) and (ii) the surface energy or interfacial tension of the substrate.

Of great relevance to the present description is the concept of *surface free energy*, the origin of which is related to the fact that surface molecules have unexploited (and available) attractive forces on one side. In the case of a liquid, the presence of attractive forces, which are directed toward the interior, results in minimization of the surface area, and since it is the tensile force that is responsible for the surface area reduction, one refers to the force as *surface tension* (related to the interface between two phases). In this context, a useful concept is that of a *contact angle* (related to the edge of the boundary between two phases and a third phase, i.e., the solid–liquid–vapor junction). The surface tension and the contact angle are closely interrelated, and, in principle, one can derive the surface tension from the contact angle measurements.

The commonly employed techniques for measuring the contact angle are the *sessile drop* (i.e., the static contact angle), the *Wilhelmy plate*, and the *du Nouy ring* methods (see, e.g., Petrie, 2000).

A quantitative description of the wetting phenomena, using the behavior of a drop of liquid on a solid surface, can be carried out by employing the so-called Young's equation. Specifically, the contact angle made by an adhesive drop on a substrate surface can be defined as the angle formed between the tangent along the profile of the uncured adhesive and the substrate at the solid–liquid interface (see, e.g., Neumann and Spelt, 1996), as depicted in Figure 3.1. This property is a manifestation of the difference in the surface energy of the adhesive and the substrate (adherend). In practice, some finite angle θ and a finite contact line are established, signifying that (at that line) the solid, the liquid, and the vapor phases are in contact. For a perfectly uniform substrate, Young's equation can be used to determine the (static) contact angle from

$$\gamma_{LV} \cos \theta = \gamma_{SV} - \gamma_{SL} \qquad (3.8)$$

where the contact angle θ is related to the interfacial tension at the liquid–vapor (γ_{LV}), solid–vapor (γ_{SV}), and solid–liquid (γ_{SL}) boundaries. Thus, in principle, this equation provides a relationship between the contact angle and the

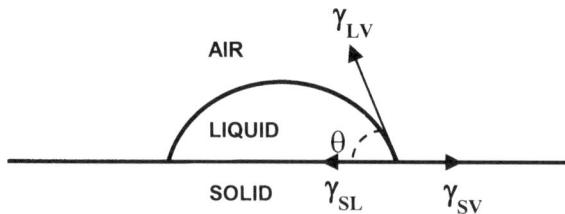

Figure 3.1 Schematic representation of the contact angle on a smooth surface and its surface tension (surface free energy) components.

interfacial energies. The actual systems, however, typically have heterogeneous surfaces both in shape and in chemistry. Thus, the incorporation of a correction factor into this equation is required in order to account for regions of different interfacial tensions.

For a curved liquid surface, there is (in equilibrium) an associated pressure difference across the interface. The degree of curvature is a function of the interfacial tension of the liquid–air interface and the pressure difference across the interface, and these can be related by using the Young–Laplace equation:

$$\Delta P = \gamma \left(\frac{1}{R_1} + \frac{1}{R_2} \right) \tag{3.9}$$

where ΔP is the pressure difference across the interface, γ is the interfacial tension, and R_1 and R_2 are the two principal radii of curvature. If the gravitational force and the interfacial tension at the solid–liquid boundary are similar, then the shape of the liquid is described as a sessile drop (see Figure 3.2). Measurement of the contact angle and the interfacial tensions can be made using the axisymmetric drop-shape analysis method (see, e.g., Neumann and Spelt, 1996). This technique essentially involves imaging a sessile drop of the adhesive on a substrate surface (see Figure 3.2) and deriving a fit to a Laplacian profile from numerical integration of Equation (3.9). Thus, in principle, the value of the surface energy of the liquid can be determined from the best-fit profile to the experimental data. The value of the surface energy of the substrate can be calculated from the contact angle at the interface between the adhesive and the substrate.

The *dynamic contact angle* measurement can be performed by using the Wilhelmy plate method. In this case, a thin plate is suspended vertically above a liquid, and the force of a liquid on the plate is measured as it passes through its surface.

It should be noted that the parameters involved in this description (i.e., θ, γ_{LV}, γ_{SV}, and γ_{SL}) relate to the macroscopic behavior at a distance between about 0.1 and 1 μm from the contact line, and θ in this case should be referred to as the *macroscopic contact angle*, which can be measured, in practice, by using optical microscopy techniques. In principle, however, at the contact line within a distance on the order of a nanometer to tens of nanometers from the contact line, the shape of the drop can be somewhat different (see Figure 3.3), indicating certain ambiguity and the need for inquiry into the reliability of using measurements such as the macroscopic contact angle for quantifying the wetting phenomena.

The surface energy of the adhesive must exceed that of the adherend by at least 10 mJ/m^2 for adequate wetting of the surface. The surface energy of a liquid or a solid in millijoules per square meter is dimensionally identical to the related value of the surface tension in millinewtons per meter (or dynes per centimeter). For the example of a liquid, these values are numerically the same.

(a)

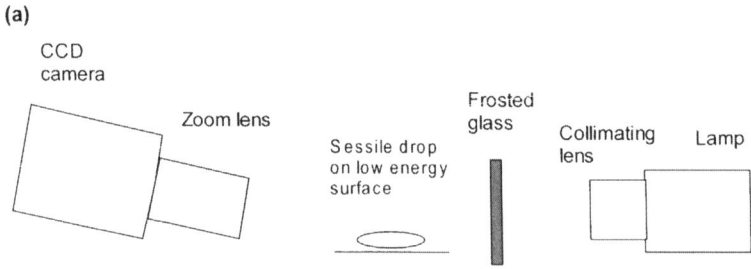

(b)

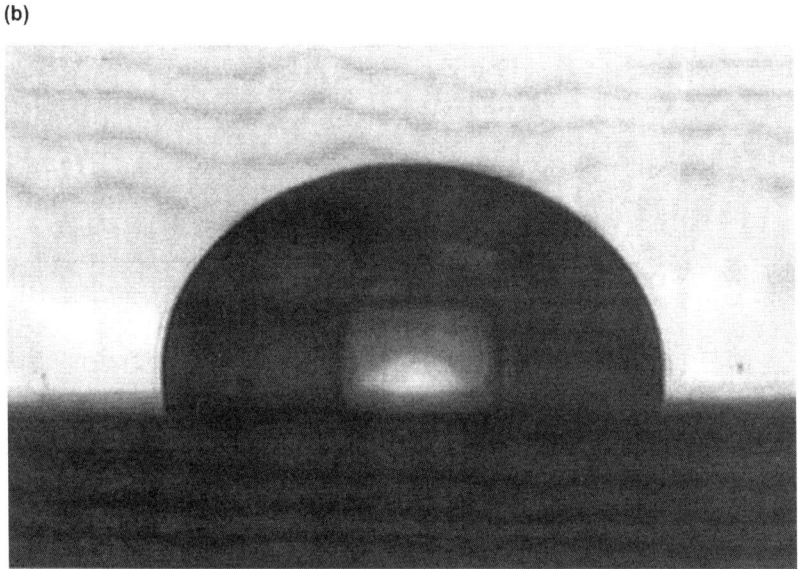

Figure 3.2 (a) Experimental setup for drop-shape analysis and (b) image of the sessile drop. (Note that the sessile drop method of measuring surface tension involves determining the shape of the drop resting at equilibrium on a surface, i.e., it does not wet a surface.)

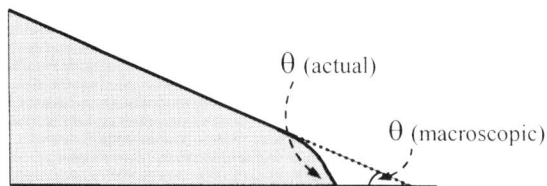

Figure 3.3 Schematic representation of the distinction between the actual contact angle and the macroscopic contact angle.

For this reason, the terms are often interchanged in the literature and the units for the surface energy are used for values of the surface tension and vice versa. However, for solids, these quantities can be different.

Adequate wetting of the substrate by the adhesive will be achieved when the surface energy of the solid phase is greater than that of the liquid phase. The surface energies of typical substrate materials are listed in Table 3.3.

The surface energy values of adhesive materials can vary in the range between about 25 and 50 mJ/m^2. However, the surface energy of the substrate is reduced if contaminants are adsorbed on the substrate.

In the case of a nonuniform substrate, the shape of cavities on the surface and the surface energy will determine the degree of wetting by the adhesive. An increase in adhesive strength might be expected through mechanical keying of rough surfaces, thereby increasing the total interfacial area. This will require penetration of the adhesive into cavities on the surface, which, however, might be prevented by the presence of trapped air providing a back pressure on the adhesive droplets.

The adhesive strength is reduced by the presence of contaminants on the surface of the substrate (typically, these are weakly adsorbed organic molecules or condensed moisture). This necessitates the use of various surface treatment techniques for adhesion improvement. The surface can be treated chemically, using solvents, oxidants, strong acids, or bases. Alternative methods of treatment for metallic and ceramic surfaces involve excimer or CO_2 lasers or bombardment with high-energy ions, resulting in ablation of surface contaminants (in addition, this can lead to increased surface roughness arising from the removal of the outer layers of the substrate). One can also employ surface adhesion promoters such as silanes.

In the context of wetting and spreading issues in adhesion bonding, it should also be emphasized that, in practical cases involving porous substrates (e.g., in the cases of ceramic materials), additional concerns related to the effect of

Table 3.3

Surface Energies of Selected Materials

Material	Surface energy (mJ/m^2)
Polypropylene	24
Polyethylene	31
Polyvinylchloride (PVC)	40
Glass	170
Methanol	23
Typical adhesive materials	~25–50
Distilled water	72

porosity arise. In this case, one has to consider two competing processes related to viscous spreading of an adhesive on a substrate and its absorption by a capillary suction into the pores of the substrate. In principle, the surface of the porous substrate can be modified in order to prevent the infiltration of an adhesive into the pores. This may be realized by using an intermediate polymer film deposited on the surface of a porous substrate in order to obtain a uniform layer on the porous substrate.

3.4. THEORIES OF ADHESION

In general, several theories of adhesion have been advanced in order to describe the different types of adhesive bonding. It should be noted that no single theory is capable of explaining all types of adhesive interaction and bonding, although it is certain that chemical interfacial interactions make a major contribution to the strength of various types of adhesive bonds. The principal adhesion theories include (i) adsorption theory, (ii) chemical bonding, (iii) diffusion theory, (iv) electrostatic attraction, and (v) mechanical interlocking. It should be emphasized that, depending on the specific surfaces involved, different mechanisms and/or combinations of mechanisms may be involved in adhesive bonding. In addition, it should always be remembered that various defects and contamination might significantly affect adhesion characteristics. In other words, these adhesion theories apply to perfect surfaces.

The formation of an adhesive bond is accompanied by the establishment of an intermediary zone at the interface between adherend and adhesive. The physical and chemical properties of the adhesive in such an interphase region may substantially differ from the properties of the adhesive away from the contact regions. It is important to emphasize that the composition of this interphase region determines the strength and durability and environmental degradation of the adhesive bond.

Adsorption theory relates adhesion to the interatomic or intermolecular (depending on the character of the adherend) attractive forces between the adhesive and the adherend. Such attractive forces between the adhesive and the adherend at the interface may be due to, for example, van der Waals bonding. This adhesion mechanism has found substantial experimental support. According to this mechanism, the wetting of the adherend by the adhesive is a key factor in determining the strength of the adhesive bond, and given that an adhesive is required to wet the surface, this mechanism contributes to the adhesion strength in each case. (Note that, as mentioned above, in order to wet the surface of the adherend by the adhesive, the surface tension of the adhesive must be lower than that of the adherend.)

Chemical bonding (or chemical reaction theory) proposes that the properties of the adhesive bond result from interfacial forces or chemical bonding. Thus, chemical bonding is a factor, provided there is an establishment of covalent, ionic, hydrogen, or van der Waals bonds between the adhesive and the surface. According to this mechanism, that is, adsorption and surface reaction, bonding occurs when adhesive molecules, adsorbed on a surface, react chemically with it. This mechanism differs, in principle, from simple adsorption (because of the chemical reaction), but some consider chemical reaction to be part of a complete adsorption process and, thus, not a separate adhesion mechanism.

Diffusion theory is operational, for example, in the case of bonding between polymers having mutual solubility, which results in the interdiffusion between the chains of molecules in polymers at elevated temperatures above their glass transition temperature T_g. In this case, the strength of the adhesive bond is considered to be associated with the extent of polymer interdiffusion across the interface. Thus, this mechanism mostly relates to the bonding between two polymers, but it is not expected to be a factor in, for example, the bonding between a polymer and a metal due to the absence of solubility of the polymers with metals.

Electrostatic attraction suggests that electrostatic forces develop at an interface between materials with differing electronic band structures. These forces are due to the formation of an electrical double layer of separated charges of opposite sign at the interface between dissimilar materials and are considered to be a contributing factor in the resistance to severance of the adhesive and the adherend. Thus, the adhesive bond in this case can be related to a charged capacitor, with the adhesive strength described in terms of the force that is required to separate the plates of a capacitor. Adhesives and adherends containing polar molecules or permanent dipoles are likely to form such electrostatic bonding.

Mechanical interlocking involves the surface roughness of the adherend, which provides both a greater number of interlocking sites and a larger surface area for the bond. Such interlocking may take place in the case of adhesive flowing into pores in the adherend surface or around protrusions on the surface. Note, however, that, under stress, excessive roughness may cause detachment (from the bulk) of the adhesive in the voids, resulting in a weaker bond and subsequent failure. Although experiments do demonstrate improved adhesion associated with the roughening process (mechanical or chemical), it is thought that this mechanism is only partially responsible for the adhesion in cases involving an adherend surface having characteristics that are conducive to mechanical interlocking. The mechanical interlocking mechanism must have a certain effect on the strength of an adhesive bond, based on the fact that adhesive bonds formed on roughened surfaces are typically stronger (with all other factors being equal), as compared to well-polished surfaces.

It should be mentioned that, in addition to the above mechanisms, *weak boundary layer theory* relates weak adhesive strength and interfacial failure as being due to the presence of weak boundary layers. This theory explains the failure of adhesive bonding as being not at the adhesion interface, but within the adhesive or adherend. This implies the formation of a weak boundary layer, which contains impurities and experiences undesirable chemical reactions around the interface.

In summary, it is certain that no single mechanism could explain all the various cases of an adhesive joint and its strength, which most probably depends on a variety of properties, including those related to mechanical character of the joint itself and its capabilities related to load distribution. With several different mechanisms for describing the adhesive bond strength, and with various possible failure mechanisms in practical applications, it is very difficult to reduce the description of adhesive bonding in terms of simple concepts or algebra. Thus, in practice, the main issue may not be the determination of a single mechanism responsible for the adhesive bonding, but, more likely, the identification of a dominant mechanism (or mechanisms) and properties that are accountable for the adhesive joint.

3.5. TYPES OF JOINT GEOMETRIES

Careful joint design can be as important as the appropriate choice of adhesive. One of the primary objectives in joint design is ensuring the distribution of the load throughout the bonded area and, thus, minimizing stress concentrations and maximizing the contact area.

In principle, adhesively bonded joints may experience various types of stresses, such as tensile, compressive, peel, and/or shear, which may also be present in various combinations. Thus, the design of the joint geometry is of great importance, considering that (i) adhesively bonded joints are relatively strong under tension, compression, and shear loading; and (ii) such joints are not as strong in cleavage and peel. It should be emphasized that the details of the joint design, related to its geometric characteristics and the manner of the transmission of the applied load, have a major effect on the mechanical properties of the adhesively bonded joint. The types of stresses that are typically present in adhesive joints are illustrated in Figure 3.4.

As a main characteristic, adhesive joint designs of higher quality are those having larger contact areas between the components to be joined (see Figure 3.5). In this context, *lap joint* designs are advantageous for bonding thin rigid cross-sectional parts. Note that in lap joints the bonded parts are slightly offset, resulting in the development of peel and cleavage forces in the presence of the load. However, by employing an offset lap joint, such forces can be reduced.

TENSILE STRESS

COMPRESSIVE STRESS

SHEAR STRESS

CLEAVAGE STRESS

PEEL STRESS

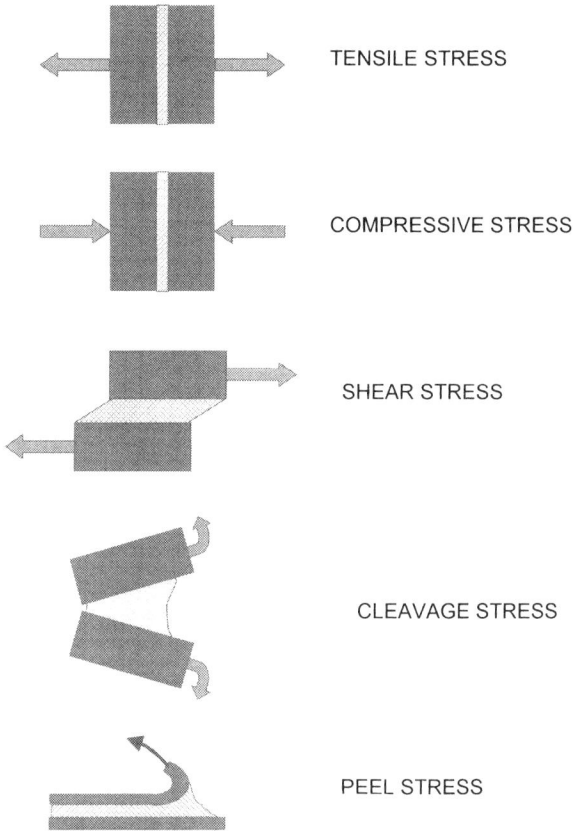

Figure 3.4 Types of stresses that are typically present in adhesive joints.

3.6. SURFACE MODIFICATION TECHNIQUES FOR ADHESION IMPROVEMENT

Surface treatment of an adherend is one of the most important issues related to adhesive bonding, with insufficient or inappropriate treatment being one of the most probable causes of adhesive bond failure. Indeed, given that adhesive bonding is fundamentally related to surface attachment, the properties and condition (such as the presence of any contamination) of the adherend surface are of paramount importance. This typically involves cleaning of surfaces that are frequently contaminated with dirt, oil, moisture, or other impurities. In some cases of metals, such as aluminum, an oxide layer formed on the surface provides a suitable surface for an adhesive. However, other cases, such as glass, necessitate specific surface treatments for appropriate adhesive bonding.

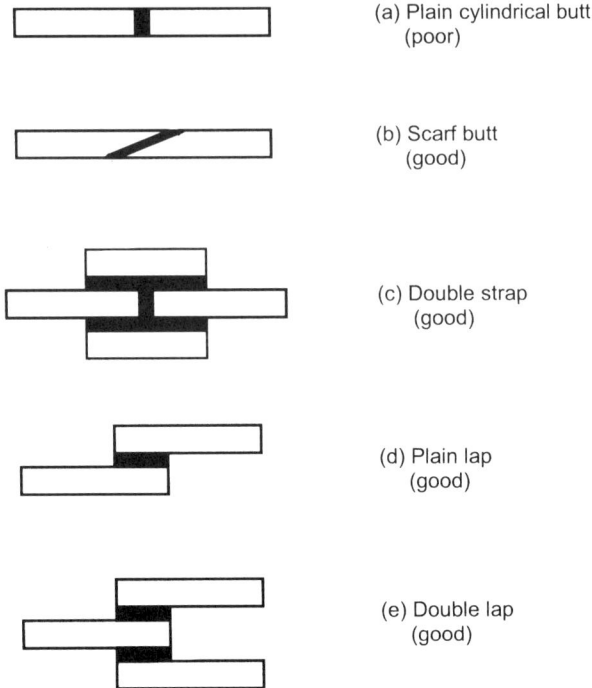

(a) Plain cylindrical butt
(poor)

(b) Scarf butt
(good)

(c) Double strap
(good)

(d) Plain lap
(good)

(e) Double lap
(good)

Figure 3.5 Selected types of adhesive joint designs. (Note that an increase in bond area results in improved bond joint.)

As mentioned above, the adhesive strength is typically reduced by the presence, on the substrate surface, of contaminants, including, for example, weakly adsorbed organic molecules or condensed moisture. In general, the surface can be treated chemically, using solvents, oxidants, strong acids, or bases. Alternative methods of treatment for metallic and ceramic surfaces involve excimer or CO_2 lasers or bombardment with high-energy ions, resulting in ablation of surface contaminants. In addition, this can lead to increased surface roughness arising from the removal of the outer layers of the substrate. Typical contaminants present may also include oils and weak oxide layers on metals, and fluorocarbons and silicones on polymers. For example, even low levels (i.e., less than 1 at%) of fluorine contamination on the surface can result in a significant reduction in the adhesion strength of polyimide materials to substrates such as silicon wafers.

There are several surface modification techniques and other methods to improve adhesion of organic coatings (see, e.g., Mittal, 1996, 2000). These include *plasma cleaning*; *flame and corona discharge*; *UV radiation*, including the *laser surface treatment* technique; *electron-beam* and *ion-beam treatment* techniques; *silane adhesion promotion*; *mechanical abrasion*; *solvent cleaning*, followed by *wet chemical etching*; and the application of *chemical primers*. Among these techniques, treatment with high-energy radiation (such as the UV laser radiation, electron-beam, and plasma methods) is based on employing reactive energetic species (i.e., photons, electrons, ions, and free radicals) that interact with the material surface and modify its chemistry and/or morphology.

Plasma treatment to improve adhesion involves exposing (in a vacuum chamber) the material's surface to plasma consisting of various excited species, which are produced by excitation of the gas molecules, subjected to an electric field, by free electrons colliding with neutral gas molecules and dissociating them into various reactive species. Consequently, the interaction of these excited species with the material's surface results in its chemical and physical modification. In general, the process related to plasma processing of a specific material is mainly controlled by the chemistry of the reactions between the reactive species in the plasma and the surface. In typical applications, employing low-energy surface treatment, the plasma treatment is limited to the near-surface region (a few monolayers deep), and it does not affect the material's bulk properties. Plasma treatment typically employs various gases (depending on the application) such as air, oxygen, argon, nitrogen, helium, nitrous oxide, methane, carbon dioxide, and tetrafluoromethane. The ensuing modification of the surface due to plasma treatment depends on the specific gases used and the composition of the material's surface. In this context, one should consider the following questions: How quickly after treatment is the surface recontaminated, and how does one handle the surface after treatment?

Flame and corona discharge methods to improve adhesion are both based on exciting gas molecules (and producing species such as ions, electrons, or neutrals) that subsequently collide with the surface and cause its chemical modification. These treatments are typically performed in air and are, thus, conducive to a wide variety of industrial applications.

In *UV/ozone treatment*, the material's surface is exposed to both UV light and ozone, which results in the incorporation of an increased amount of oxygen functional groups. This method is suitable for the treatment of surfaces of three-dimensional objects.

Chemical priming can facilitate improved adhesion by applying a chemically specific layer on the surface. Some of the examples of primers include acrylates and polyurethanes. There is no universal primer for all materials, however, and thus specific materials are required for different applications.

The addition of *adhesion promoters* can improve the adhesive strength between a polymeric material and an inorganic surface. For example, in the case of a glass substrate, silane agents can be used for this purpose. This additive will, in addition, provide a barrier to moisture permeation along the

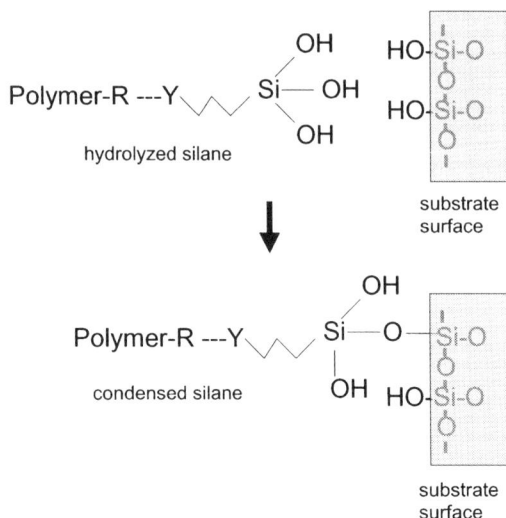

Figure 3.6 The silane agent forms a strong chemical bond to the substrate, improving the wettability of the surface and the strength of the adhesive bond.

adhesive interface, which is a common cause of adhesive bond failure. The silane molecules are able to form strong chemical bonds with the glass substrate through hydrolyzable alkoxy groups. In addition, the same molecules contain polarizable side groups capable of forming a physical bond with the polymer (see Figure 3.6). Note that, typically, the adhesion promoters such as silane need to be heated to temperatures greater than 60 °C to become activated and promote adhesion.

3.7. SUMMARY

Adhesive bonding is employed for many applications (e.g., bonded structures, electronics assembly and packaging, photonics assembly and packaging, medical device manufacturing, composite materials, automotive technology, and paints). An adhesive is defined as a material capable of holding two other materials together in a practical manner by surface attachment that resists separation. Synthetic organic adhesives are typically composed of *polymers* (i.e., *macromolecules* formed by the linking of many simpler molecules known as *monomers*). Polymers are typically flexible, and they are able to spread and interact on the surface of the substrate material (this is highly important in practical applications of adhesives). The relevant physical properties of polymers (e.g., melting point, viscosity, solubility, and tensile strength) are determined by the strength of the intermolecular forces, the molecular weight, the polymer structure, and the flexibility of the polymer molecule. The formation of the polymer, that is, *polymerization* (i.e., a chemical reaction resulting in the linking of molecules of monomers and forming of polymers through chain growth) can be achieved by a photochemical reaction, resulting in the conversion of the liquid monomer into a solid polymer. The process of the conversion of an adhesive from a liquid to a solid state is termed *curing*, which can be realized in one manner by radiation curing (e.g., optical beam, electron beam, X-rays, microwaves).

Most adhesives are composed of several constituents performing various functions. These constituents may include adhesive base or binder, solvents, initiator, accelerators, inhibitors, or retardants, fillers, and carriers.

The main advantages of joining components with adhesives include the following: (i) Adhesives provide a continuous bond, with uniform distribution of load over wider areas; (ii) adhesives can seal against moisture penetration and thus prevent corrosion; (iii) adhesives can bond irregularly shaped surfaces relatively easily; and (iv) adhesives can make a bond more resilient to stress. In addition, adhesives are advantageous in both cost and convenience as compared to mechanical means, and they are especially better suited to the assembly of fiber-optic components and structures. The disadvantages of adhesive bonding include (i) a more limited (as compared to mechanical methods) service temperature range, (ii) the dependence of the adhesive strength on the condition of the surface of the adherend, and (iii) the possible deterioration of adhesive bonding in the presence of water and/or water vapor.

A variety of variables have to be considered in order to understand the specific adhesive bond. No single model or simple relationships containing constants and materials parameters could explain all types of adhesive interaction and bonding. The major factors that influence adhesion are (i) interactions at different length scales, (ii) surface morphology, and (iii) deformation processes at different length scales. The performance of the adhesive bonding is mainly determined by (i) the physical and chemical properties of the adhesive, (ii) the type of adherend, (iii) the nature of the surface treatment, (iv) the wettability of the surface, and (v) the details of joint design.

CHAPTER 4

Types of Adhesives

CONTENTS

4.1. INTRODUCTION

In general, there is a wide selection of adhesives for various applications, requiring different performance characteristics and specific processing conditions, dependent on a particular material or system being cured or on a specific application. An industrial guide, *Adhesives, Sealants and Coatings for the Electronics Industry* (Flick, 1992), contains descriptions of a wide variety (over 2500) of adhesives, sealants, and coatings. The current generation of adhesives with improved performance has found new areas of applications and they are even considered as replacements for such methods as welding and soldering.

This book is concerned with synthetic organic adhesives, which can be produced in large quantities (with consistently uniform properties) and which can be modified and combined in a myriad of ways in order to obtain the customized characteristics for specific applications.

The trend in the development of new adhesives is toward reducing (or eliminating, if possible) the use of solvents and substituting them with waterborne, hot-melt, or completely solid materials. For some applications, there is also a need for adhesives that can tolerate harsh environments, such as wide temperature fluctuations and corrosive surroundings.

As mentioned above, synthetic organic adhesives are typically composed of polymers. The formation of the polymer, that is, *polymerization* (defined as a chemical reaction resulting in the linking of molecules of monomers and forming of polymers through *chain growth*, i.e., addition polymerization, or *step growth*, i.e., condensation polymerization), can take place (i) during the curing process (outlined below), resulting in simultaneous polymerization and formation of adhesive bond; or (ii) prior to the material being applied as an adhesive.

Some relevant concepts of polymers and polymerization will be discussed in the following sections.

4.2. INTRODUCTION TO POLYMERS AND POLYMERIZATION

As mentioned above, *polymers* are *macromolecules* that are formed by the linking of many simpler molecules known as *monomers*. In the context of adhesion, the main advantageous properties of polymers include their flexibility and their ability to spread and interact on the surface of the substrate material. The latter is of vital importance in the practical application of adhesives. The basic physical properties of polymers are mainly determined by the strength of the intermolecular forces, the molecular weight, the polymer structure, and the flexibility of the polymer molecule. This section outlines briefly the basic properties of polymers.

In general, polymers have found a wide range of applications in various fields, including adhesives, coatings, composites, various devices (e.g., electronic, optical, and biomedical), and a wide range of precursors for ceramics.

As a general characteristic, a polymer has a repeating structure (resulting in long chainlike molecules), which is typically based on a carbon backbone. The main advantages of this type of material include low cost, relatively easy processing at low temperatures, low weight, and corrosion resistance.

As mentioned above, the formation of the polymer, *polymerization*, is defined as a chemical reaction resulting in the linking of molecules of monomers and forming of polymers.

The two types of polymerization reactions are (i) *chain-reaction* (or addition) polymerization and (ii) *step-reaction* (or condensation) polymerization. The chain-reaction (or addition) polymerization process involves three steps and two chemical reactants, that is, a monomer (or a link in a polymer chain) and a catalyst. The monomers typically contain at least one carbon–carbon double bond. An example of the catalyst in a chain-reaction process is a free radical, which contains an electron forming a covalent bond with an electron on another molecule. The three steps in chain-reaction polymerization include (i) initiation (i.e., a double-bonded carbon monomer reacts with a free-radical catalyst), (ii) propagation (i.e., a repetitive action resulting in the formation of a physical chain of the polymer), and (iii) termination (e.g., through combination, i.e., two growing chains joining and forming a single chain). The step-reaction (or condensation) polymerization process relates to polymer formation through stepwise intermolecular reaction involving at least two monomer species.

In general, some basic arrangements of polymers include (i) *linear* (i.e., polymers consist of a single long continuous chain), (ii) *branched* (i.e., consisting of a main chain of molecules with shorter molecular chains that are branched from the main chain), and (iii) *cross-linked* (i.e., polymers with valence bonds forming between discrete chain molecules). Structurally, polymer segments can assume crystalline or amorphous forms. In crystalline polymers, molecules are arranged in an ordered manner (i.e., in polymer crystals, the chains are lined up in an ordered manner), whereas in amorphous polymers the chains are tangled up in various ways.

The main characteristics that determine the physical properties of polymers are (i) the molecular weight, (ii) the structural order of the polymer, (iii) the strength of the intermolecular forces, and (iv) the flexibility of the polymer molecule.

An important property of a polymer is the glass transition temperature T_g, which relates to the transformation from a rigid material to a material that has the characteristics of a rubber. Typically, at the glass transition temperature, the mechanical properties of polymers change from those corresponding to relatively hard and even brittle materials (below T_g) to relatively soft and flexible materials (above T_g). Above the glass transition temperature, there is a large change in the coefficient of thermal expansion. The glass transition temperature depends on the structural properties of the material, in particular, the chain flexibility. The flexibility is determined by the ability of the molecular units to rotate and vibrate; decreased flexibility will inhibit rotational motion as the temperature is increased. Flexibility can be reduced by increased chain length, branching, or cross-linking, or by the introduction of bulky groups to the polymer chain. A somewhat ambiguous term for a polymer is the melting point, which typically occurs over a range of a few tens of degrees Celsius (in this range, the polymer's viscosity gradually changes from that corresponding to a solid state to that of a liquid). It should be noted that, in principle, one should distinguish between the melting temperature and the glass transition temperature. Melting relates to a transition occurring in crystalline polymers when the polymer chains are transformed into a liquid, whereas the glass transition relates to amorphous polymers. Note that, because of the nature of their structure, amorphous polymers do not have a specific melting temperature (T_m). In principle, since crystalline polymers in fact contain a certain amorphous fraction, the polymer sample can have both T_g and T_m, with the crystalline fraction undergoing melting only and the amorphous fraction undergoing the glass transition only.

The glass transition temperature can be modified by incorporating a *plasticizer*, that is, a component added to a polymer in order to enhance flow, flexibility, and deformation. This is accomplished by incorporating a small molecule between the polymer chains, which results in increasing the distance between the chains that makes it easier for them to move relative to each other (thus, resulting in lowering of the T_g of the polymer).

Typically, at low temperatures, a polymeric material possesses elastic characteristics and, at higher temperatures, viscous (liquid) properties. For intermediate conditions, a polymer can display a combination of the mechanical characteristics of elastic and viscous behavior; the material is referred to as viscoelastic. The elastic component of the polymer deformation is instantaneous; that is, the strain is displayed the instant the stress is applied or released and independent of time. This fraction of the total deformation is completely recovered when the external force is released. The viscous component to an applied stress is not instantaneous; that is, the resulting strain is delayed and dependent on time. This fraction of the total deformation is irreversible and not completely recovered after the stress is released. Therefore, viscoelastic behavior is characterized by an instantaneous elongation of a material following application of a force, followed by an additional viscous elongation dependent on time. The rate of increase in the strain can determine whether the deformation is elastic or inelastic. The time-dependent increase in strain when the stress level is maintained constant is referred to as viscoelastic creep. This behavior can be exhibited by polymeric materials at room temperature and with values for the applied stress below the yield strength. Understanding the creep behavior of polymeric materials is vital in the applications related to manufacturing of photonics components that require high long-term stability.

4.2.1. MECHANICAL PROPERTIES

The mechanical properties of polymers depend most significantly on temperature, with polymers exhibiting brittle (glasslike) behavior at low temperatures and a rubber-like behavior at high temperatures.

The *strength* of polymers can be related to various cases, such as (i) *tensile* strength (i.e., the case when the material is stretched or under tension), (ii) *compressional* strength (i.e., strength related to the case when material is compressed), (iii) *flexural* strength (related to bending of the material), (iv) *torsional* strength (related to the twist), and (v) *impact* strength.

Elongation is a type of deformation related to a change in shape as a result of applied stress. For example, in the case of tensile stress, the material can deform by getting longer; thus, this is referred to as elongation, measured as a relative quantity (or percentage of elongation), that is, as the length of the sample after stretching, divided by the original length of the sample, and multiplied by 100. In measurements, one can determine the so-called *ultimate elongation* (i.e., the amount of possible stretch of the sample prior to breaking) and *elastic elongation* (i.e., the percentage of elongation that can be reached without permanent deformation of the sample).

The modulus of elasticity (or Young's modulus) E of the material is expressed as $E = \sigma/\varepsilon$, where σ represents *stress*, that is, the force (F) normalized by the cross-sectional area (A) of the material ($\sigma = F/A$); and ε is *strain*, that is, the change in length of the material normalized by the initial length. (Young's moduli for selected materials are listed in Table 4.1.)

In general, the mechanical properties of polymers depend significantly on factors such as (i) rate of strain, (ii) temperature, and (iii) environmental conditions. Thus, the stress–strain characteristics may exhibit various types of behavior such as brittle, plastic, and elastic (or rubber-like). The modulus of polymers (see Table 4.1) is typically substantially smaller as compared to other types of materials, but elongation can be up to 1000% in some cases.

Table 4.1

Young's Moduli for Selected Materials (1 GPa $= 100,000$ N/cm^2)

Material	Modulus (GPa)
Polymers	
Rubbers	0.01–0.1
Polyethylene	0.2–0.7
Epoxies	2–3
Nylon	2–4
Polyesters	1–5
Polyimides	3–5
Metals	
Aluminum (Al)	69
Steel	190–200
Tungsten (W)	406
Ceramics, glasses, and semiconductors	
Soda-lime glass (Na$_2$O–SiO$_2$)	69
Silica glass (SiO$_2$)	94
Silicon (Si)	107
Aluminum oxide (Al$_2$O$_3$)	390
Silicon carbide (SiC)	450
Diamond (C)	1000

It should also be noted that although the stress–strain curve is typically straight, in some cases of polymers (e.g., plastics) the curve, which is straight only at the initial stage, saturates to a specific value. This implies that the slope of the curve varies as a function of stress.

An additional important parameter for characterizing materials is *toughness*, represented by the area underneath the stress–strain curve. This property is a measure of the energy that can be absorbed by the sample prior to breaking. The difference between strength and toughness is that strength is related to the force that is required to break a sample, whereas toughness relates to the amount of energy that is required to break it. It should be emphasized that the high strength of the material does not necessarily imply correspondingly high toughness. The material can be strong but not tough. In addition, a strong material that can deform much may break; that is, the material is brittle. However, from a stress–strain curve consideration, there is a certain combination of the properties of the materials that may exhibit both high strength and high toughness. This is related to the fact that deformation generally permits the dissipation of energy, whereas the absence of deformation inhibits energy dissipation and results in the breaking of a sample.

An important characteristic of polymers is that of *viscoelasticity*, that is, a material's property associated with a combined elastic and viscous behavior. This property is related to the fact that, while at low temperatures amorphous polymers typically deform elastically, at high temperatures they exhibit viscous behavior. In contrast, at intermediate temperatures, polymers exhibit a (viscoelastic) characteristic that is similar to a rubbery solid.

One of the important mechanical properties of polymers is the fracture strength, which is typically considerably lower than that of metals. Fracture, which usually involves breaking of covalent bonds in the polymer chains, originates with cracks at various imperfections. Understanding the factors related to the fracture mechanics of adhesive joints is of great importance in manufacturing photonics components and systems.

The two important polymeric materials, which are extensively employed in industry, are *plastics* and *elastomers*. Plastics, which are typically processed by forming or molding into a specific shape, can be divided into two main categories (related to their behavior when heated), that is, *thermoplastic* (e.g., polyethylene) and *thermosetting* (e.g., epoxy) polymers, determined by their structure and chemical bonding.

Thermoplastic polymers can repeatedly become elastic or melt when heated (and, thus, can be shaped by flow into specific objects by molding or extrusion), and they return to their hardened state by cooling through the specific temperature range characteristic of a specific material. Thermoplastic polymers (typically with linear molecules containing carbon) are synthesized by addition or condensation polymerization. This results in strong covalent bonds within the chains and weaker secondary (van der Waals) bonds between the chains (see Section 3.2). The presence of such secondary bonds between the chains makes thermoplastic polymers moldable at high temperatures.

Thermosetting polymers undergo an irreversible chemical cross-linking reaction from liquid to solid (i.e., they harden permanently); these polymers are typically harder and more brittle (but more dimensionally stable) than thermoplastic polymers. These do not soften under heat and cannot be remolded. In thermosetting polymers, strong covalent bonds hold different chains, which may be bonded either directly to each other or through other molecules, allowing the polymer to withstand softening upon heating. These polymers have high thermal and dimensional stability, as well as high rigidity.

Elastomers (e.g., rubbers) can be typically deformed elastically to a large extent (under the application of a tensile force) and can revert to their original form with the release of the force. Some important mechanical properties (in the context of the present discussion) are listed in Table 4.2.

4.2.2. FORMULATIONS OF ADHESIVES

In general, the formulations employed by manufacturers in the production of adhesives are withheld from the end user as a trade secret. Typically, the adhesives are identified by a series of numbers and letters that represent a product code for the actual chemical content. However, the functionality of the active ingredient responsible for the chemical activity of the material is usually specified. In most cases, adhesives are based on *acrylate*, *epoxide*, or *thiolene* chemistries (see Figure 4.1). These functional groups demonstrate the most common mechanisms for polymerization reactions. The acrylate and epoxide photoinitiated reactions proceed by a chain mechanism involving, respectively, free-radical and cationic addition of molecular units to form the polymer network. The free radicals or cations are formed by photolyzing photoinitiators

Table 4.2

Typical Ranges of Selected Mechanical Properties of Synthetic Polymers

Polymer	Young's modulus E (GPa)	Ductility (%)	Tensile strength (MPa)
Elastomers	0.01–0.1	300–2000	5–30
Thermoplastics	0.17–1.1	100–1200	8–30
Thermosets	2.4–4.4	30–300	50–100

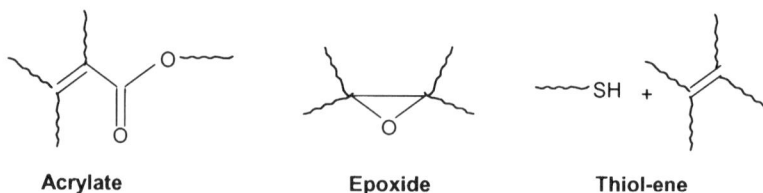

Acrylate   **Epoxide**   **Thiol-ene**

Figure 4.1 Chemical formulas for the acrylate, epoxide, and thiolene functional groups. In these formulas, not all the elements present in bonding configurations are shown, since there is an unlimited number of possibilities of those, and they are also often of a proprietary nature.

with appropriate wavelength and energy of light. In contrast, the thiolene reaction involves a step-growth addition that is propagated by a free-radical chain-transfer process (Fouassier and Rabek, 1993). Unlike the reactions of acrylates and epoxides, thiolene polymerization is based on a stoichiometric reaction of *thiol* to *ene*.

The active ingredient in an uncured resin provided by the adhesive manufacturer could itself be a polymeric material (known as an *oligomer*) with a low molecular weight. The oligomer could be derived from a partial cure of the resin or from the formation of a prepolymer. Some common resins are polyester, polyether, polybutadiene, epoxy, and polyurethane oligomers that are functionalized with an acrylate group. Alternatively, the resin can contain a monomeric species where the functional group is attached to a small molecular species (typically, a saturated hydrocarbon or a cycloaliphatic molecule). In most cases, the monomer or oligomer has more than a single functional group. This enables cross-linking of the polymer chains during the curing process. In order to modify the viscosity or the rate of the polymerization, a monomeric material can be mixed with an oligomer in the uncured resin.

The epoxide, acrylate, and thiolene chemistries illustrate the commonly encountered mechanisms for photopolymerization reactions. The epoxide and acrylate molecules react via an addition-chain mechanism in a similar fashion, despite the different nature of the intermediate species involving cations and free radicals, respectively. The photocured epoxy materials have a different chemistry than the traditional two-part epoxy resin that cures at room temperature. The traditional epoxy system involves the addition of a nucleophilic reagent (i.e., the hardener), such as an amine, to the epoxide material (i.e., the resin), as shown in Figure 4.2. Photocured epoxy materials contain a photoinitiator, usually an onium salt or an organometallic derivative. Thus, the components can be mixed and supplied in one container. Photolysis of the initiator will generate a cation that can promote the polymerization reaction (see Figure 4.3). Each addition of a molecular unit is accompanied by the formation of a cationic species. This allows propagation of the reaction in a chain process. The polymerization mechanism, shown in Figure 4.3, is illustrative of the chain-reaction mechanism.

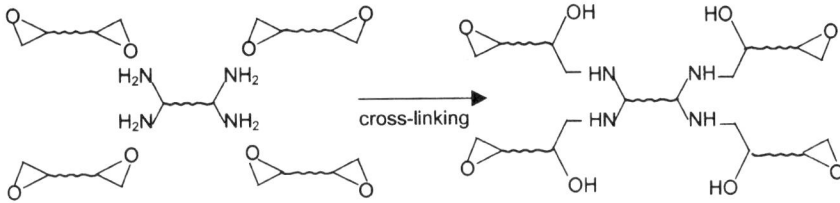

Figure 4.2 Polymerization of a traditional epoxide material. The amine acts as a cross-linking agent for the bifunctional epoxide (oligomer) molecules.

Resins can include an initiator, similar to the photoactivated species described above, where the ionic or free-radical reagent is generated at elevated temperatures. This type of initiator (called a thermal initiator) can be used in conjunction with a photoinitiator to provide a dual-cure material.

The photopolymerization reaction, shown in Figure 4.3, is nonstoichiometric. Polymerization proceeds until all the material is consumed or a nonpropagating species is produced. For acrylate materials, the propagation steps are sensitive to ambient (or dissolved) oxygen, which can react with the free-radical intermediate to produce a nonpropagating species. Therefore, in practice, an atmosphere that does not contain oxygen has to be used to cure these materials. Thiolene chemistry also involves free-radical intermediates; however, the mechanism avoids oxygen inhibition as a result of thiol being an extremely effective chain-transfer agent. In this case, the polymerization process is propagated despite the reaction of the intermediates with oxygen.

The examples of the polymerization reactions described above are those encountered for the majority of adhesive resins. However, there are other cases involving novel materials, such as anionic photoinitiators that can be used to cure epoxy materials.

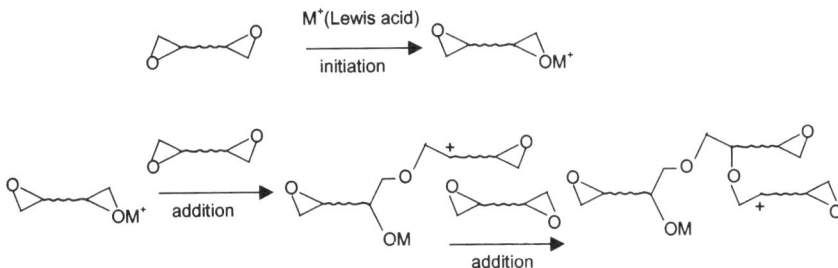

Figure 4.3 Mechanism for the UV-induced polymerization of an epoxide material. The reaction proceeds by a chain process involving cationic intermediates. A similar mechanism is exhibited by acrylate materials involving free-radical intermediates.

4.2.3. INTRODUCTION TO PHOTOPOLYMERIZATION

As mentioned above, there is no universal adhesive that is suited to all applications, which typically require different performance characteristics and specific processing conditions that are related to a particular material or system being cured or a specific application. Thus, judicious choices of a specific adhesive for a given application and of an appropriate curing method are of vital importance. As mentioned above, cure of an adhesive is a process of its conversion from a liquid to a solid state, accompanied by a physical or chemical modification to the adhesive. Changing a liquid adhesive into a solid state can be accomplished, for example, by light curing or polymerization; during cure, the properties of a thermosetting resin are irreversibly modified by chemical reaction (i.e., condensation, ring closure, or addition); cure may be realized by the addition of cross-linking agents, with or without catalyst, and with or without heat; cure may also take place by addition in cases such as epoxy resins. The critical considerations for suitable selection of an adhesive include its compatibility with the backing material and the operating temperatures at which the adhesive bond is anticipated to function.

For photonics applications, we can distinguish among three methods of adhesive curing. These include:

(a) Curing with light
(b) Curing with heat
(c) Dual light and heat curing

The chemical mechanism for photopolymerization reactions utilizing thermal energy or UV light is a chain reaction. The cleavage of a weak chemical bond in an initiator molecule leads to the formation of a highly reactive intermediate; this is achieved by either thermal activation of critical vibrational modes or electronic excitation to a dissociative state following absorption of a photon. The quantum yield for photopolymerization reactions is very large; a single intermediate generated from an initiator molecule can lead to the formation of thousands of chemical bonds between monomer molecules. The initiator is consumed in the first step of the polymerization reaction. However, the reactive sites are propagated along the polymer chain as each molecular unit is added. In this context, knowledge of reaction kinetics during photopolymerization is of great importance (e.g., Elliott and Bowman, 1999; Goodner and Bowman, 1999). These reactions can be subdivided into categories according to the functional groups present in the molecular units of the polymer (called monomers) and the electronic structure of the intermediates. For example, the polymeric materials formed from monomers containing the acrylate group are typically generated through a mechanism involving radical intermediates. In contrast, those formed from small molecules containing the epoxy group usually involve cationic inter-mediates. A cross-linked polymer network is generated when the monomer con-tains more than one of these functional groups. In this case, chemical bonds are formed between the polymer chains. The polymer network is described as a gel when the polymer chains extend throughout the material with a substan-tial increase in viscosity. When the physical forces of attraction between the polymer chains are sufficient, vitrification of the material will occur.

There are two principal classes of polymeric materials that arise from dif-ferent densities of chemical cross-linking between the molecular chains. Ther-moplastic materials contain either none or very few cross-links. As mentioned above, these materials can be softened by heating or hardened by cooling; this process is reversible. Thermosetting materials contain considerably more cross-linking bonds (as many as 50% of the monomer units). This is usu-ally achieved by heating the material to increase the flexibility of the mol-ecular chain, allowing active sites on different chains to align and chemical bonds to form. Although thermosetting polymers will initially soften when they are heated, cross-linking is an irreversible process and the material will become permanently hard. As mentioned above, thermosetting polymers are gen-erally harder, stronger, more brittle, and have better dimensional stability than thermoplastics.

The degree of cross-linking can vary between polymeric materials. Conse-quently, they can exhibit a wide range of different mechanical properties. How-ever, they typically exhibit greater plastic elongation but lower modulus and tensile strength than metals and ceramics. In addition, the mechanical properties are much more sensitive to temperature changes (sometimes in the vicinity of ambient conditions) and the rate at which a stress or strain is applied. Higher temperatures can result in decreased brittleness, greater plastic deformation or ductility, and decreased values for the elastic modulus and tensile strength.

One of the most efficient methods for producing very rapid reactions in polymers is that of laser-assisted processing (Decker, 1990). For example, from kinetic studies, the photopolymerization of acrylic systems, employing a pulsed or continuous-wave (cw) ultraviolet laser, was shown to occur almost instan-taneously, as recorded by real-time infrared spectroscopy in the millisecond time scale (Decker, 1990). Similar studies also demonstrated the power of time-resolved IR spectroscopy in analyzing the kinetics of UV-induced photopolymer-ization reactions in polymer systems (Scherzer and Decker, 1999). It was also demonstrated that such ultraviolet laser-assisted processing provides an efficient method for realizing rapid (i.e., a few milliseconds) curing of photosensitive resins (Decker, 1999).

UV light curing, which is the most widespread type of light-induced curing of adhesives, can be employed with either continuous-wave or pulsed wave irra-diation and is commonly employed in the microelectronics, medical, fiber-optic assembly, and printing industries. Pulsed UV light is a rapid curing method that allows control of cure at low average power and low temperatures. It is important that the spectral output (i.e., the intensity of light at each wavelength over the whole wavelength range emitted by the lamp) is matched with the absorption characteristics of the photoinitiator. Some important wavelength regions include 250 nm, which is associated with surface cure; 320 and 365 nm, where wavelengths greater than 350 nm improve depth cure; and the range between 400 and 500 nm, which employs visible photoinitiators to absorb light and to provide even greater depth of cure. This can be increased by employing combined cure systems that include such additional cure mechanisms as heat and activators. In general, in order to match a light source to the UV-curable adhesive, one must determine (i) the photoinitiator system that is incorporated in the adhesive and (ii) the transmission characteristics of the substrate.

Curing with heat typically requires accurate control of temperature and cure time, which is dependent on the adhesive and the type of bond required.

One of the most important steps in curing of adhesives is controlling various temperature profiles during the curing steps or the light output for photocuring. In typical applications in the fiber-optic communication industry, small amounts of adhesive are dispensed to cover small discrete areas on the work pieces, which are first aligned and subsequently held in position with jigs or fixtures, followed by the whole assembly being heated to accomplish the cure. For improved manufacturing efficiency, batch processing is employed in large thermal ovens. Nevertheless, the overall efficiency still remains low due to the fact that such an oven-based curing process constitutes a major restriction in the overall manufacturing process. Thus, substantial advantages and improved manufacturing efficiency can be realized by employing a curing method that renders an *in-line* and *in situ* processing step that can directly follow or precede other processing or fabrication steps. Such an in-line and *in situ* processing step cannot be realized by incorporating furnace heat treatments. However, such a processing step can be realized by using accelerated curing, employing the delivery and continuous real-time control of focusable optical radiation (e.g., UV, visible, or infrared radiation) to localized regions of materials only. An example of such an automated assembly system is described in Section 5.3.

To summarize, the formation of a polymer, that is, *polymerization*, can be achieved by a photoinduced reaction, leading to the conversion of a liquid monomer into a solid polymer. The process of converting an adhesive from a liquid to a solid state is termed *curing*. Optical radiation curing employs ultraviolet (UV), visible, and infrared (IR) photons. The main advantages of polymers (which are important for the formation of an adequate adhesive bond) include their flexibility, their ability to spread and interact on a material's surface, and their strength.

4.3. CLASSIFICATION CATEGORIES

In general applications, adhesives can be classified in several different ways. Adhesives can be divided into categories, which are described below, related to their chemical composition, or properties, or curing methods, or function, or reaction method. Note that these different categories of adhesives have a certain overlap; however, for convenience and completeness, we will use various classification systems, so that the description of a specific adhesive in the literature may, in principle, appear in different classification sections.

Classification by function distinguishes adhesives as being (i) structural and (ii) nonstructural. The former are adhesives with high strength and durability, and their basic function is related to their ability to hold structures together under elevated loads and without deformation. The function of nonstructural adhesives is simply to hold materials (of light weight) in position; these types of adhesives are typically degraded by extensive exposure to environmental factors (e.g., temperature and chemicals) and they creep under relatively reduced loads.

Classification by chemical composition distinguishes adhesives among (i) *thermoplastic adhesives* (these usually have reduced heat and creep resistance and are typically used for low-load conditions), (ii) *thermosetting adhesives* (these typically exhibit good creep resistance and are used for high-load conditions), (iii) *elastomeric adhesives* (these are typically polymeric resins with increased toughness and elongation, and they are especially suitable for bonding to materials with dissimilar expansion coefficients), and (iv) *hybrid adhesives* (these are typically blends of the three previous categories offering combinations of preferred characteristics).

Classification based on reaction method is relatively broad and distinguishes adhesives on the basis of their solidification mode, such as (i) chemical reaction (e.g., reaction with radiation, heat, surface catalyst, or hardener component), (ii) cooling from a melt, and (iii) loss of solvent.

Classification based on physical form includes those adhesives that include (i) one part solution (in liquid form), (ii) one part solventless (liquid or paste), or (iii) several parts solventless (liquid or paste), solid adhesive (e.g., tape, film, powder). These all have their advantages and disadvantages.

Classification by type of adherend or specific application refers to the adherend (i.e., the material to which the adhesive is being bonded), and it distinguishes between, for example, wood adhesives and metal adhesives. Also, the adhesives can be classified by the specific application form (e.g., spraying, brushing, extrusion, syringe application).

In some applications related to photonics and microelectronics, it is also convenient to describe adhesives based on the following categories.

Ultraviolet (UV)-cured adhesives (employed mainly in the electronics, photonics, automotive, and medical fields) are characterized by relatively low process temperatures and relatively high polymerization rates (from several seconds to about 60 seconds). In electronics applications, these are mainly based on *acrylic* and *urethane* chemistries and some light-curing *epoxies* and *silicones*. These adhesives consist of (i) monomers; (ii) various agents and modifiers (e.g., wetting agents, stabilizers, fillers); and (iii) photoinitiators, which are only activated with exposure to light of a specific wavelength and intensity. The energetic free radicals, generated by the photoinitiators, start the formation of monomer chains, and after several steps, the cross-linked polymer chains are completely reacted, that is, cured. Combining the ultraviolet with a visible light cure is advantageous, since ultraviolet curing of thick materials can lead to a cure gradient, and by using an ultraviolet/visible adhesive, it is possible to attain a more consistent cure profile. These adhesives, often containing various fillers (e.g., borosilicate glass or alumina), have improved characteristics related to lower shrinkage and electrical and thermal conductivities. (Note that the *filler* in this case is defined as a material incorporated into the adhesive to facilitate the modification of its properties, e.g., physical, electrical, thermal, and mechanical.) Among this type of adhesive, *aerobic adhesives* offer highly suitable bonding for optical assembly applications, with the advantages of faster cure, low stress, and high-strength bonds. It should be noted that combined ultraviolet/heat curing adhesives are also available for various applications.

Optical adhesives, in general terms, can be defined as adhesives used in various applications for bonding different photonics components and structures. In a narrower sense (in specific applications), optical adhesives are those that are characterized by a refractive index adapted to the component material of the optical assembly. Such adhesives, developed for various optical assemblies, have refractive index and molecular structure designed (by optimizing formulations) to provide controllable refractive indices in the range between about 1.40 and 1.60 and high transmission in selected wavelength regions (e.g., visible, 1300–1600 nm, etc.). In a wider sense, however, optical adhesives are characterized by a broad range of appropriate optical, mechanical, and thermal properties, which make possible the use of adhesive bonding in the joining of various optical components and systems. In such (general) applications, considerations of great importance (and a selection of an optical adhesive for a specific application) relate to such characteristics as (i) appropriate refractive index, (ii) high optical clarity, (iii) low shrinkage during cure, (iv) insignificant relative thermal movement between bonded components, (v) low stress on bonded components, (vi) high glass transition temperature (T_g), and (vii) low outgassing.

Electrically conductive adhesives are distinguished between two basic types of conductive adhesives. These are (i) *isotropically conductive adhesives* (ICAs) and (ii) *anisotropically conductive adhesives* (ACAs). For ICA applications, Ag is typically employed as a filler material, whereas, for ACA applications, polymer-based metal-plated spheres or Ni fillers are typically used.

Pressure-sensitive adhesives (PSAs) include adhesive tapes and films, with the capability of holding materials together as a result of the application of pressure at room temperature. These PSA adhesives include thermoplastic elastomers, rubbers, polyacrylates, and silicones. These types of adhesives, however, are not suited for continuous loading.

Anaerobic adhesives harden in the presence of metal and in the absence of oxygen (or air). These are typically based on acrylic resins.

Structural adhesives provide high strength and durability; elevated load-transferring capability; and resistance to solvents, heat, and fatigue. This type of adhesive includes (i) *epoxies* (having high strength and temperature resistance), (ii) *acrylics* (very versatile adhesives with capabilities of fast curing and bonding to oily surfaces), (iii) *polyurethanes* (these are flexible and resistant to fatigue and have excellent peeling characteristics), (iv) *anaerobics* (suitable for bonding cylindrical shapes), (v) *silicones* (these provide sealing capability and are durable in the open air), and (vi) *cyanoacrylates* (poor resistance to moisture and temperature but have fast bonding capability to plastic and rubber). All the above types of adhesives can be also be modified in order to obtain adhesives with various properties.

Hot-melt adhesives are used as nonstructural adhesives. These are typically thermoplastic resins that melt at high temperatures and are applied to the surface as hot liquids. These include polyamides, polyesters, elastomers, and rubber. Their main disadvantage is related to their inferior strength at high temperatures, but modifying them with reactive urethanes and polyethylenes has led to improved adhesives of this type. For improved performance at higher temperatures, structural hot-melt adhesives have been prepared with enhanced peel adhesion and relatively higher heat capability.

Epoxy adhesives consist of (i) an epoxy resin and (ii) a hardener, and since there is a wide variety of both of these, this type of adhesive offers great flexibility for various formulations. These adhesives typically form very strong and durable bonds with a wide range of materials, and they are usually obtainable as one-part, two-part, or film forms.

Toughened adhesives (e.g., acrylic and epoxy adhesives) typically incorporate a dispersed (chemically attached, but physically separate) rubber phase, resulting in an adhesive with better fracture and impact resistance.

4.4. ADHESIVE SELECTION CRITERIA

The selection of an adhesive for a specific application depends on such uncured and cured properties as viscosity, pot life (i.e., time the adhesive remains workable), cure time, postcure strength and hardness, chemistry (which determines those materials that can be adhered), shrinkage, T_g, moisture resistance, and density.

Some important aspects of the adhesive selection criteria include the following:

(a) Identification of critical characteristics of the interface between the optical components and the mount

(b) Substrate materials used

(c) Bond area and geometry

(d) Required bond strength

(e) Required bond line thickness

(f) Evaluation of contact stresses present due to mounting forces

(g) Requirement for the bond to be electrically or thermally conductive

(h) Requirement for the bond to be highly transparent and have suitable refractive index

(i) Low shrinkage after cure to minimize any movement resulting in misalignment

(j) Ability to be stable for long periods of time under severe environmental conditions, including high temperature and humidity

(k) Thermal expansion coefficient mismatch between the adhesive and adherend

In addition to these requirements, there are also those related to the dispensing (this is often not considered appropriately, leading to poor bond performance and poor repeatability) and curing considerations, such as the viscosity of the adhesive, the cure speed, and the permissible handling time.

As mentioned above, there is a wide selection of adhesives for various applications, requiring different performance characteristics and specific processing conditions, dependent on the particular material or system being cured or on a specific application, including applications related to assembling various photonics components, structures, and devices. Adhesive manufacturers list thousands of adhesives designed for very specific applications, and their information databases provide very specific sets of properties for each of their formulations (which are of a proprietary nature). Thus, although it is impractical to describe all of these (even if their formulations were known) in the limited format of this book, the selected adhesives manufacturer could certainly direct one toward the most suitable adhesive for a desired application. (Note again that synthetic organic adhesives can also be modified and combined in a myriad of ways in order to attain the customized characteristics for specific applications.)

Some general properties of selected adhesives are given in Table 4.3.

It should be noted that in recent years there have also been a number of efforts aimed at developing so-called *expert systems* for selecting adhesives for various applications involving conflicting design criteria in aerospace hardware design (Darmody and Chadwick, 1987; Darmody and Schneemann, 1989), in the microelectronics industry (Estes, 1991; Su et al., 1993; Derebail et al., 1994), and in the fabrication of military hybrid circuits (Estes, 1986).

Table 4.3

Typical Properties of Selected Adhesives

	Service temperature (°C)	Advantages	Disadvantages
Epoxies	−40–150	High bond strength Good chemical and temperature resistance Versatile Wide variety of formulations Good gap filling	Short pot life Requires special equipment for dispensing Low peel strength High exotherm Slow fixturing Toxicity
Acrylics	−40–100	Versatile adhesives with capabilities of fast curing Good chemical and temperature resistance Good impact, peel, and shear	Primer required Flammable Toxicity
Polyurethanes	−50–120	Good resistance to fatigue Versatile Good peel strength Good toughness	Moisture sensitive Poor resistance to temperature Short pot life Toxicity
Cyanoacrylates	−30–100	Fast bonding capability to plastic and rubber High tensile strength Easy dispensing Good adhesion Excellent pot life	Poor resistance to moisture and temperature Poor gap filling Cured adhesive is brittle Poor peel strength
Anaerobics	−55–150	Suitable for bonding coaxial joints Versatile Easy dispensing High strength Excellent pot life Good solvent resistance	Typically brittle Usually requires primers
Silicones	−70–260	Flexible Good for bonding glass to other substrates Good moisture and temperature resistance Variety of viscosities	Relatively low strength Short shelf life Corrosive Slow curing Limited solvent resistance
Polyimides	−45–300	High-temperature applications	Demanding processing during assembly

4.5. DISPENSING METHODS

The developments in photonics packaging (including automated assembly practices) necessitate improved methods of adhesive dispensing. In general, one of the difficult issues is to develop a repeatable means of dispensing fluids with a wide range of properties (especially a range of viscosities). From dispensing adhesives by employing a handheld syringe, current developments are directed toward automated dispensing for volume production. In this context, one should note that the dispensing techniques must be compatible with the component assembly rates. In addition, the continuous miniaturization of optoelectronic components and devices requires new processing methods for microdispensing of adhesives.

Manual adhesive dispensing is a difficult process to control, as it is susceptible to operator error, and it may substantially delay the assembly process. In addition, the viscosity of an adhesive may change as the result of environmental temperature fluctuations and with various deposition variables related to placement accuracy, volume, and dispenser configuration. The contactless adhesive dispenser has great advantages in applying adhesives, since it eliminates the vertical movement and decreases the dispensing time, and, consequently, increases the throughput.

In general, the various types of adhesive dispensing tools available in practice include hand syringes, micropens, ink jets, dispense jets, and auger valves.

The characteristics of manual and air-powered dispensers include their ability to dispense microdeposits in a consistent manner and their use with two-part epoxy and UV adhesive. Air-powered dispensing, which is a low-cost method requiring little maintenance, is the most widely employed method of dispensing for photonics applications at present, but it is somewhat operator dependent.

In jet dispensing, adhesive is fed into the chamber and is heated to a uniform temperature (making it difficult to use with thermally initiated adhesives). A specific design allows the adhesive to fill the void as the ball is retracted. With the ball returning, the adhesive is jetted through the nozzle to the substrate, where it develops an adhesive dot. In this method, multiple shots can facilitate an increasing dot size.

Some of the limitations of ink jets include (i) the difficulty in cleaning, (ii) the fact that it can only be used with low-viscosity fluids, and (iii) the inability to be used with filled materials such as silver epoxy.

Among these methods, dispensing methods employing jets, pumps, and valves have a wide viscosity range and provide good repeatability.

Present efforts in photonics assembly are directed toward automation, which also include in-line automated dispensing systems.

It should be noted that the continuing trend toward miniaturization necessitates a precise method for depositing very small volumes of material, such as adhesives. One of the promising techniques for small-volume dispensing is the microvalve technique, which provides a dispensing system with accurate control of the x, y, and z positions of the dispensing needle tip, as well as with improved repeatability of the deposited material. In this case, the valve includes a rotary auger pump and an encoder for precision. Thus, such a valve is designed to dispense material based on an encoder that accurately controls the rotation of the auger pump during the delivery of the adhesive through the needle. Such a system also provides great dispenser repeatability. The dispensing needle in such systems is fabricated from stainless steel. In order to reduce surface tension between the adhesive and the needle, the tip has a conical cut in the direction of adhesive flow. Such characteristics of the needle provide for improved adhesive flow that helps to avoid clogging.

4.6. SUMMARY

There is a wide selection of adhesives available for various applications. Synthetic organic adhesives, which are typically composed of polymers, can be produced in large quantities, and they can be modified and combined in a myriad of ways in order to obtain the customized characteristics for specific applications. The formation of the polymer, that is, *polymerization*, can take place (i) during the curing process, resulting in simultaneous polymerization and formation of an adhesive bond; or (ii) prior to the material being applied as an adhesive.

The advantageous properties of polymers (in relation to adhesion) include their flexibility and their ability to spread and interact on the surface of the substrate material. The two types of polymerization reactions are (i) *chain-reaction* (or addition) polymerization and (ii) *step-reaction* (or condensation) polymerization. Some basic arrangements of polymers include (i) *linear*, (ii) *branched*, and (iii) *cross-linked*. Structurally, polymer segments can assume crystalline or amorphous forms.

The characteristics that determine the physical properties of polymers include (i) the molecular weight, (ii) the structural order of the polymer, (iii) the strength of the intermolecular forces, and (iv) the flexibility of the polymer molecule. An important property of a polymer is the glass transition temperature T_g, which relates to the transformation from a rigid material to a material that has the characteristics of a rubber (the glass transition temperature depends on the structural properties of the material, in particular, the chain flexibility). At the glass transition temperature, the mechanical properties of polymers change from those corresponding to relatively hard materials (below T_g) to those corresponding to relatively soft and flexible materials (above T_g). The glass transition temperature can be modified by incorporating a *plasticizer*, that is, a component added to a polymer in order to enhance flow, flexibility, and deformation.

The mechanical properties of polymers depend mainly on (i) the rate of strain, (ii) the temperature, and (iii) environmental conditions. The stress–strain characteristics may exhibit various types of behavior such as brittle, plastic, and elastic. The modulus of polymers is typically substantially smaller as compared to other types of materials.

Important characteristics of polymers are viscoelasticity (i.e., a material's property associated with combined elastic and viscous behavior) and the fracture strength (which is typically considerably lower than that of metals).

The two important polymeric materials, which are extensively employed in industry, are plastics (thermoplastic and thermosetting polymers) and elastomers (which can be typically deformed elastically to a large extent and can revert to their original form).

For photonics applications, one can distinguish among three methods of adhesive curing. These include (i) curing with light, (ii) curing with heat, and (iii) dual light and heat curing. The chemical mechanism for photopolymerization reactions utilizing thermal energy or UV light is a chain reaction.

The formation of the polymer, that is, *polymerization*, can be realized by a photoinduced reaction, leading to the conversion of the liquid monomer into a solid polymer. The process of converting an adhesive from a liquid to a solid state is termed *curing*. Optical radiation curing employs ultraviolet (UV), visible, and infrared (IR) photons. The main advantages of polymers (which are important for the formation of an adequate adhesive bond) include their flexibility, their ability to spread and interact on a material's surface, and their strength.

Adhesives can be classified according to their chemical composition, or properties, or curing methods, or function, or reaction method. In some applications related to photonics and microelectronics, it is also convenient to describe adhesives based on other categories, such as ultraviolet (UV)-cured adhesives, optical adhesives, electrically conductive adhesives, pressure-sensitive adhesives, anaerobic adhesives, and epoxy adhesives.

The selection of adhesive for specific applications depends on viscosity, pot life, cure time, postcure strength and hardness, chemistry, shrinkage, T_g, moisture resistance, and density. Some principal criteria for the adhesive selection include such characteristics as (i) substrate materials used, (ii) bond area and geometry, (iii) required bond strength, (iv) required bond line thickness, (v) evaluation of contact stresses present due to mounting forces, (vi) the requirement for the bond to be electrically or thermally conductive, (vii) the requirement for the bond to be highly transparent and have suitable refractive index, (viii) low shrinkage after cure, and (ix) ability to be stable for long periods of time under severe environmental conditions, including high temperature and humidity.

CHAPTER 5

Photopolymerization Techniques

CONTENTS

5.1. INTRODUCTION

There are several types of radiation curing methods (see, e.g., Pappas, 1992). These include optical radiation (i.e., ultraviolet, visible, and infrared curing), microwave radiation curing, and electron-beam curing. The curing of adhesives using optical radiation, in the ultraviolet, visible, and infrared ranges, has become one of the most effective methods for the accelerated curing of various adhesives in applications such as bonding components in microelectronics and fiber optics for the telecommunications industry. Such curing of adhesives provides great advantages related to continuous miniaturization, manufacturing yield, and cost reduction. In addition, the combination of UV and visible light provides improved cure speeds and depths and permits a wider range of applications. It should be noted that a cure gradient might be obtained with inadequate penetration of light into the bulk of the material. In this case, the cure depth depends on (i) the wavelength of light used and (ii) the absorption properties of the adhesive material and thickness of the adhesive bond.

The initiators in light-cured adhesives can possess an absorption profile with an intense maximum in the visible, UV-A, UV-B, or UV-C regions. (The ranges for the UV are as follows: 315–380 nm for UV-A, 280–315 nm for UV-B, and 100–280 nm for UV-C.) However, typically, the breadth of the profile incorporates more than one of these regions. Consequently, the majority of light-cured adhesives can be treated with a wide range of wavelengths; therefore, an initiator with an absorption maximum centered in the UV-C region can still be treated with longer wavelengths appropriate for achieving a greater depth of cure. It is merely necessary to modify the intensity of the light source according to the absorption cross section for the wavelengths used. However, in some cases, the application of light having short wavelengths with a low penetration depth can be appropriate. The polymerization reactions that involve free-radical mechanisms (e.g., the majority of acrylic materials) are subject to oxygen inhibition, a process that will occur at the surface of the uncured liquid. As a result, a uniform cure throughout the material can be achieved using a light source that produces a much higher intensity at the surface than in the bulk. In addition, an adhesive that experiences inhibition, such as described above for acrylic materials, can be modified to contain a concentration of the initiator appropriate for different applications. For instance, a process requiring the dispensing of a thin film of adhesive will be subject to greater inhibition; consequently, a better cure might be achieved with a larger concentration of initiator (however, one must be careful that no unused radicals are left after the reaction, since this could cause corrosion in electro-optic applications).

Light-based systems are able to deliver faster cures for adhesives than traditional assembly methods (Fouassier, 1995). Efficient delivery of energy from a lamp source to the material is achieved by using either a light guide with a narrow diameter or a focusing lens. The technique is referred to as *spot curing*, and, in this case, the light energy is deposited in a localized area where the adhesive has been dispensed. The lamp source is selected to provide the appropriate wavelength range of light to cure the material.

Control of the exact dose of light delivered to the material through the measurement and setting of the intensity and duration of the exposure, in addition to the wavelength of light used, allows for customization of the curing profile for a particular application. Repeatability of the curing procedure can then be achieved using *spot-curing methods*, giving rise to consistent properties of the cured material (Hubert, 2001).

In some cases, application of a variable-intensity profile is an effective means to optimize the curing process (Fouassier, 1995). The delivered intensity of light is varied as a function of time in a series of discrete steps. Such a procedure is referred to as Step Cure™ (Hubert, 2001; see Figure 5.1). Step Cure is based

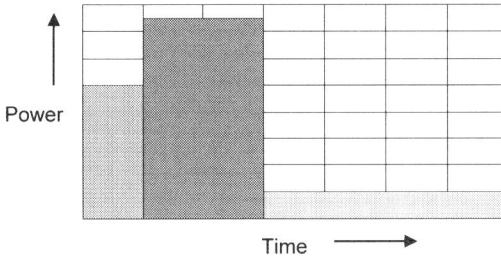

Figure 5.1 Example of Step Cure.

on controlling the intensity level during the exposure cycle, which can provide for control of the physical characteristics of the polymerized material (such as shrinkage) and ensure complete polymerization to prevent postcure material changes and outgassing. A procedure combining Step Cure with temperature change monitoring of the material would achieve greater control over the curing process. This involves incorporation of a noncontact temperature probe into a spot-curing device to monitor the progress of the highly exothermic polymerization reaction. Control can be achieved by using the output from the probe as feedback to adjust either the intensity or the duration of the exposure. This method enables a consistent level of cure to be achieved regardless of the volume or geometry of the sample. An example of Step Cure is illustrated in Figure 5.1.

In summary, optical radiation curing is typically divided among ultraviolet (UV), visible, and infrared (IR) types, corresponding to about 200–400 nm for the UV, 400–750 nm for the visible, and 750–10,000 nm for the IR. A combination of UV and visible light can also provide improved cure rates and depths and facilitate the so-called photothermal processing of materials.

In typical applications, optical radiation curing offers great advantages and improved manufacturing efficiencies, since such curing methods readily render an in-line and *in situ* processing step that can directly follow or precede other processing or fabrication steps.

It should be noted that mid-IR curing facilitates selective heating of the adhesive and fast curing rates. [The main ranges of IR radiation are *near IR* (750–1200 nm), *short-wave IR* (1200–2000 nm), *medium-wave IR* (2000–3000 nm), and *long-wave IR* (3000–10,000 nm).]

Cure depth is a critical characteristic of the curing process. Light absorption by any material depends on wavelength. Higher energy (i.e., shorter wavelength) UV is typically absorbed very near the surface region of the material, and thus, it is limited to applications to very thin layers, whereas lower energy (i.e., longer wavelength) UV penetrates farther. Some materials do not transmit UV light

well, and some have UV-blocking species that are added to avoid UV light
degradation. Thus, UV curing of thick materials can lead to a cure gradient with
the material on top being cured better than the material at the bottom.

There are certain advantages to using a combination of UV/visible/IR curing
mechanisms. Aerobic urethane encapsulants are cured with such a combination
of UV/visible/IR cures. UV and visible light provide cure (in seconds) of a
thick skin over the surface of the encapsulant. Consequently, IR cures portions
of the adhesive that were not exposed to the light (i.e., under the die edges or
underneath components).

5.2. OPTICAL RADIATION CURING

A representative system is illustrated in Figure 5.2, which presents a
schematic diagram of the integrated system (consisting of a light delivery module and a control unit) for regulated localized infrared curing of materials and

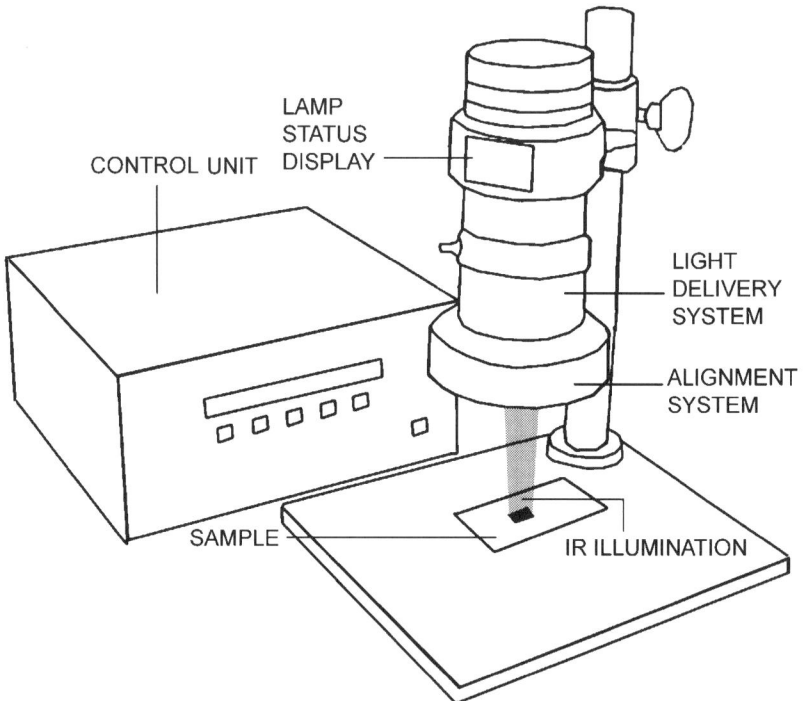

Figure 5.2   Schematic diagram of the integrated system (consisting of light delivery module and
control unit) for regulated localized infrared curing of materials and components.

components. As mentioned above, such systems offer a high level of repeatability through a closed-loop feedback system, which continuously monitors light output at the source for the corresponding adjustment of the output. This ensures the maintenance of the required intensity levels for every cure. Such microprocessor-controlled exposures with a closed-loop feedback system can also facilitate Step Cure for curing at variable and programmable conditions designed to tailor the cure profile for optimized performance. Thus, such integrated systems for highly regulated curing offer improved levels of control and repeatability, leading to improved yields.

There are great advantages to the rapid thermal curing of thermally activated adhesives, which is based on the delivery and continuous real-time control of focusable infrared radiation to localized regions of materials. Such a selective deposition of radiant energy that is immediately transferred into thermal energy results in immediate heating of only the area containing the adhesive, and it causes only limited temperature increase in the surrounding regions. In addition, this localized heating of the adhesive can be further improved by tuning of the spectral content of the radiant energy in order to match the absorption spectrum of a given adhesive material. Thus, by such a localized increase in the temperature at the area of interest only, the effective temperature can be raised to higher values as compared with the oven-curing methods.

Thus, utilization of directed infrared beams offers several advantages, including

(a) *In-line* processing step that can directly follow or precede other processing or fabrication steps

(b) Processing of materials and components of various types, such as semiconductors, polymers, and ceramics (note that different materials require substantially different thermal treatments, all of which cannot be accomplished for a given sample in the same furnace environment, but can be realized by employing real-time regulation of the intensity and duration of the infrared-beam-induced delivery of energy to selected regions of the material or structure)

(c) Cost and efficiency advantages (e.g., in the case of polymers, the curing times of several hours can be reduced to minutes, rendering the curing process as an efficient in-line curing method for the manufacturing process)

(d) Improved control over the temperature of the adhesive during the curing process (given lower thermal mass, as compared to the whole assembly, and due to localized heating, the energy deposition rate can also be modulated during the curing process, resulting in a controlled temperature profile within the adhesive during the curing process)

It should be noted that there is another method for rapid curing of adhesives that employs microwave irradiation. Although microwave radiation has a selective effect on different materials, such systems do not provide such a localized heating of a selected region as infrared irradiation.

One of the great advantages of radiation curing, that is, the flexibility in delivering a localized means for curing, is exemplified in the Figures 5.3 through 5.8 of various radiation curing systems and accessories that facilitate such flexibility. Such systems can be effectively configured and adapted for the assembly and packaging of various fiber-optic and optoelectronic components and devices. These can also be incorporated into the assembly automation practices, as outlined below.

A focusing lens, coupled to the end of a light guide, can be used to focus and collimate the light guide's output, and it allows adjustment of the spot diameter (see Figure 5.3).

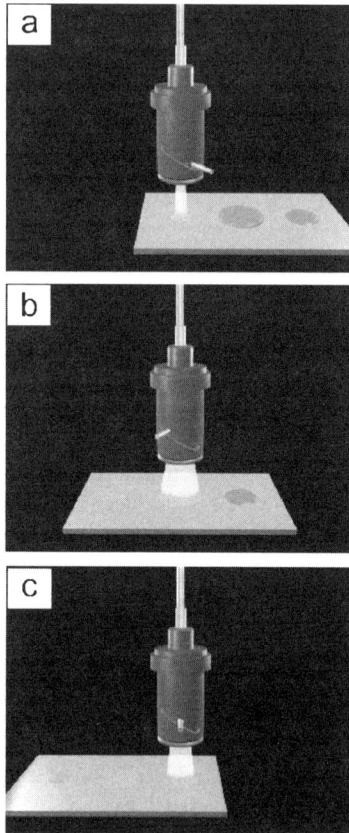

Figure 5.3 Focusing lens (coupled to the end of a light guide) used to focus and collimate the light guide's output; it also allows adjustment of the spot diameter.

In conjunction with a lens system, it is also possible to incorporate the so-called Cure Mask™, which produces various beam shapes by employing a removable Cure Mask filter (see Figure 5.4).

The *cure ring* is designed to provide 360° illumination surrounding a cylindrical area, and it thus provides a light-guide 360° curing power (see Figures 5.5 and 5.6). It can be used in a solid or hinged configuration. Such cure rings can be effectively employed for such applications as bonding cables, tubing, or systems requiring uniform 360° exposure.

The light line (see Figure 5.7) provides the capability of converting the light guide's spot of light into a focused, linear beam of curing energy. It can be configured in different ways (e.g., various dimensions) to provide a uniform linear beam of light for various applications.

The so-called *horseshoe adapter* is designed to couple to standard light guides in order to transform a circular spot into two uniform, opposed linear beams (see Figure 5.8). Such an adapter, providing equal irradiance from opposite sides, can be effectively employed in applications requiring equal illumination from two sides simultaneously. The dimensions of such an adapter can be modified for different applications. Using fiber optics for the delivery of optical radiation to a localized region, such as often required in photonics applications related to curing optical coupling material for bonding fibers to various components, necessitates using the fiber in close proximity to the work piece. The horseshoe adapter offers such a capability with no need for condensing lenses or other components for the delivery of optical radiation to a desired position only and with a desired spectral content depending on the source of radiation located outside the system. In addition, using fiber-optic delivery of radiation allows mixing of different wavelengths or changing the wavelength of delivered radiation by changing the light sources.

For curing purposes, one can also employ high-density light-emitting diode (LED) arrays that provide ultrabright, uniform illumination in the long-wavelength UV and blue regions and can be assembled in various geometries and sizes.

The key properties of adhesives that play a crucial role in various applications in the fiber-optics and microelectronics industries are, for example, (i) cure shrinkage, (ii) outgassing, (iii) moisture resistance, (iv) adhesive strength, (v) cohesive strength, (vi) T_g, (vii) rate of creep, (viii) degree of cure, and (ix) degree of cross-linking. Outgassing, moisture resistance, and rate of creep are properties that relate to the long-term stability of adhesives. These need to be investigated using appropriate accelerated testing methods. It is hoped that these efforts will lead to the development of procedures to provide tailor-made materials for different applications. The nature of the spot-curing technique ensures that these procedures will be repeatable to provide a consistent quality of the cured material.

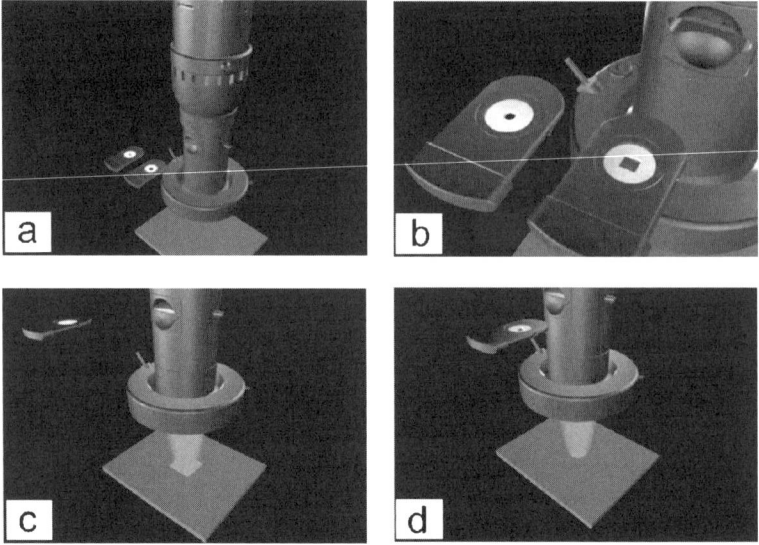

Figure 5.4 So-called Cure Mask, which (in conjunction with a lens system) enables one to produce various beam shapes by employing a removable Cure Mask filter.

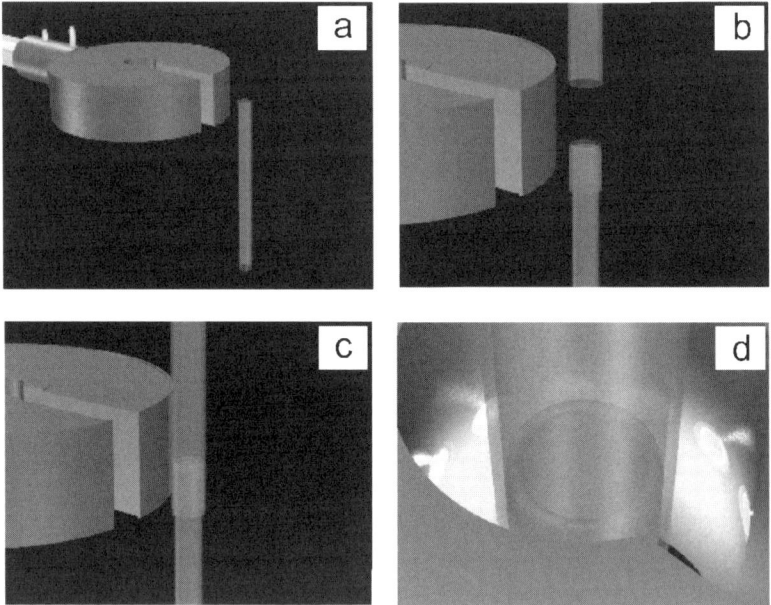

Figure 5.5 Solid configuration of the cure ring designed to provide 360° illumination surrounding a cylindrical area, thus providing a light-guide 360° curing power.

Figure 5.6 Hinged configuration of the cure ring designed to provide uniform 360° exposure.

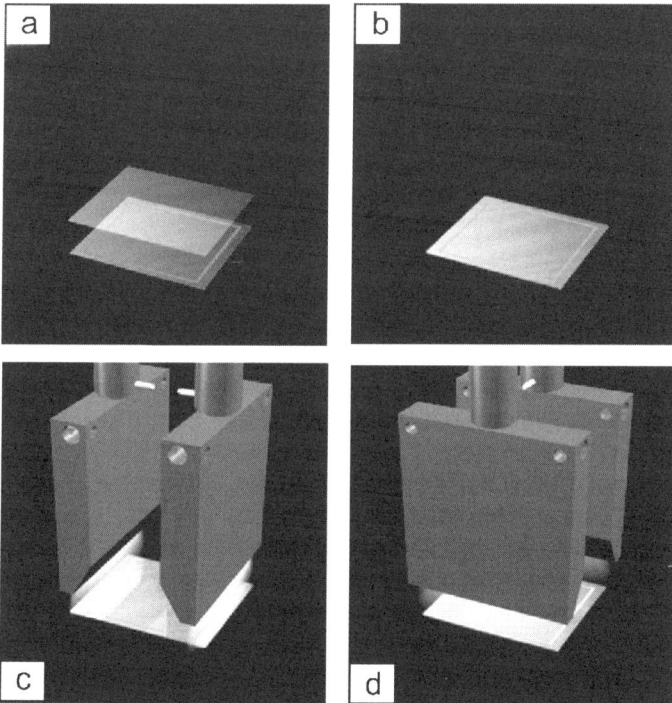

Figure 5.7 Light line providing the capability of converting the light guide's spot of light into a focused, linear beam of curing energy.

Figure 5.8 So-called horseshoe adapter designed to couple to standard light guides in order to transform a circular spot into two uniform, opposed linear beams.

It should be emphasized that:

(a) Radiation curing provides the greatest flexibility in delivering a localized means for curing.

(b) Radiation delivery systems can be effectively configured and adapted for the assembly and packaging of various fiber-optic and optoelectronic components and devices. These delivery systems are extremely important for optimizing the curing process.

(c) These can also be effectively incorporated into the assembly automation practices, as outlined below.

5.3. ASSEMBLY AUTOMATION

One of the important trends in the large-scale production capabilities of the fiber-optic industry is the development of a low-cost automated assembly process, which can be realized by employing a machine-vision-based multiaxis automated positioning system and/or by incorporating robotic "pick and place/alignment" of components. Such systems also have to incorporate in-line means for accelerated curing of adhesives during various stages of the assembly process.

The radiation curing systems and accessories, described above, can be effectively incorporated into assembly automation as well as any of the UV/visible spot-curing systems.

An example of the automated assembly process is illustrated in Figures 5.9 through 5.11.

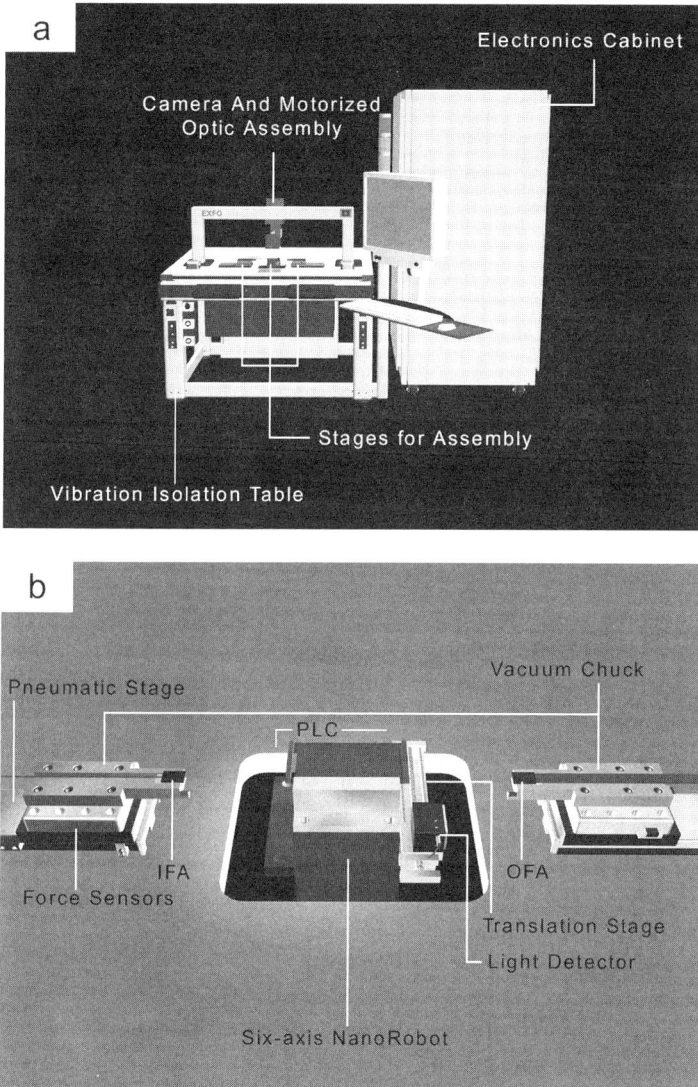

Figure 5.9 (a) Component assembly workstation for optoelectronic components (as an example, the EXFO ProAlign 5000 is shown). This is a fully integrated workstation for active alignment and assembly of photonics components. (b) Close-up view of the assembly table.

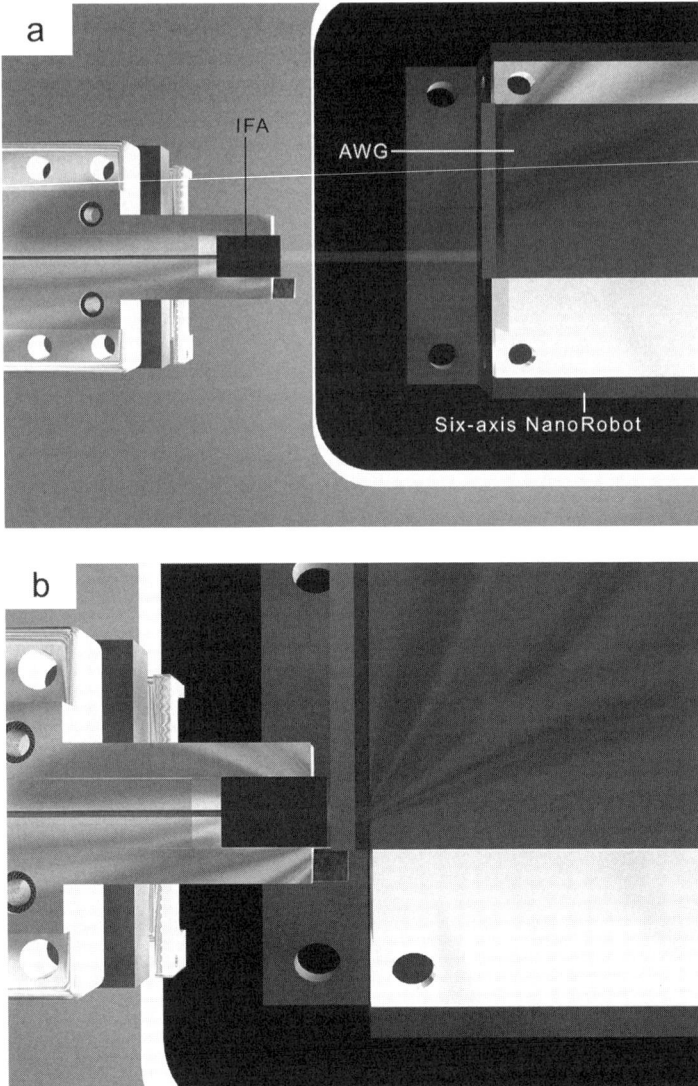

Figure 5.10 (a and b) Step 1: The input fiber array is moved into place by the pneumatic stage. The touch sensor establishes the optimum gap to begin the alignment procedure. (c and d) Step 2: The six-axis nanorobot unit carries out a search pattern in order to obtain maximum signal. The light detector moves into place to detect light and begin the alignment procedure. (e and f) Step 3: This step provides bonding of the input fiber array (IFA) to a planar light-guide circuit (PLC). (e) Dispensing (from the syringe in this case) of a UV adhesive on the part. (f) Cure of the adhesive by employing a custom light delivery system (horseshoe adapter; see Figure 5.8) that delivers UV light. (g and h) Following the completion of cure, the light detector is backed away.

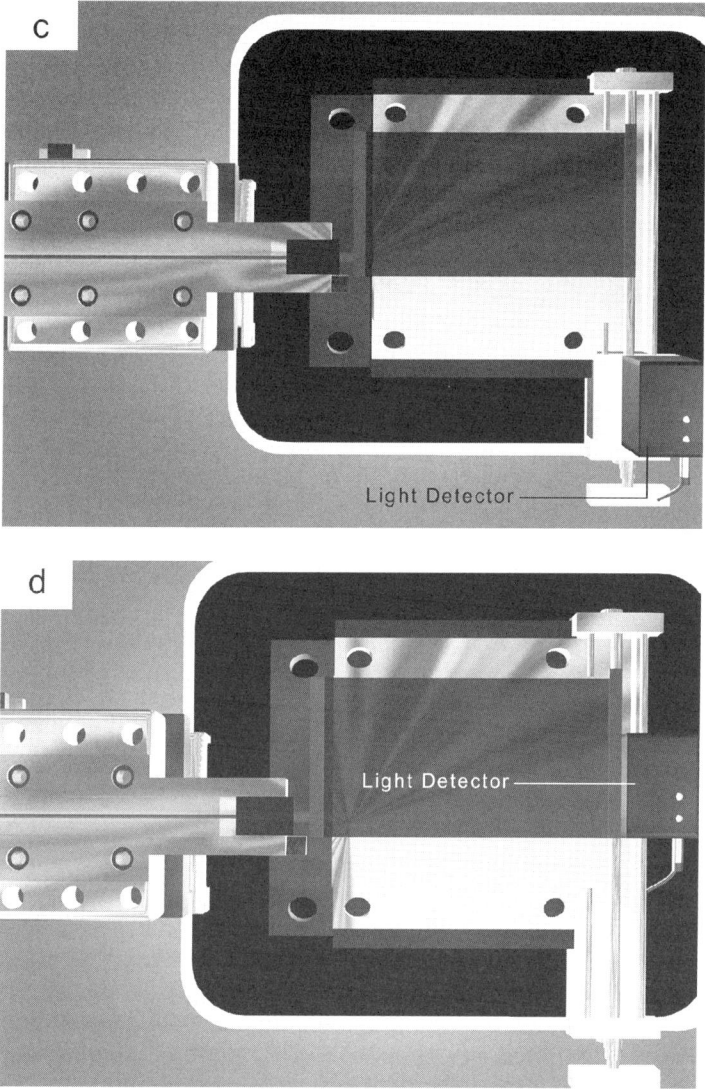

c

Light Detector

d

Light Detector

Figure 5.10 (c and d) Step 2 continued

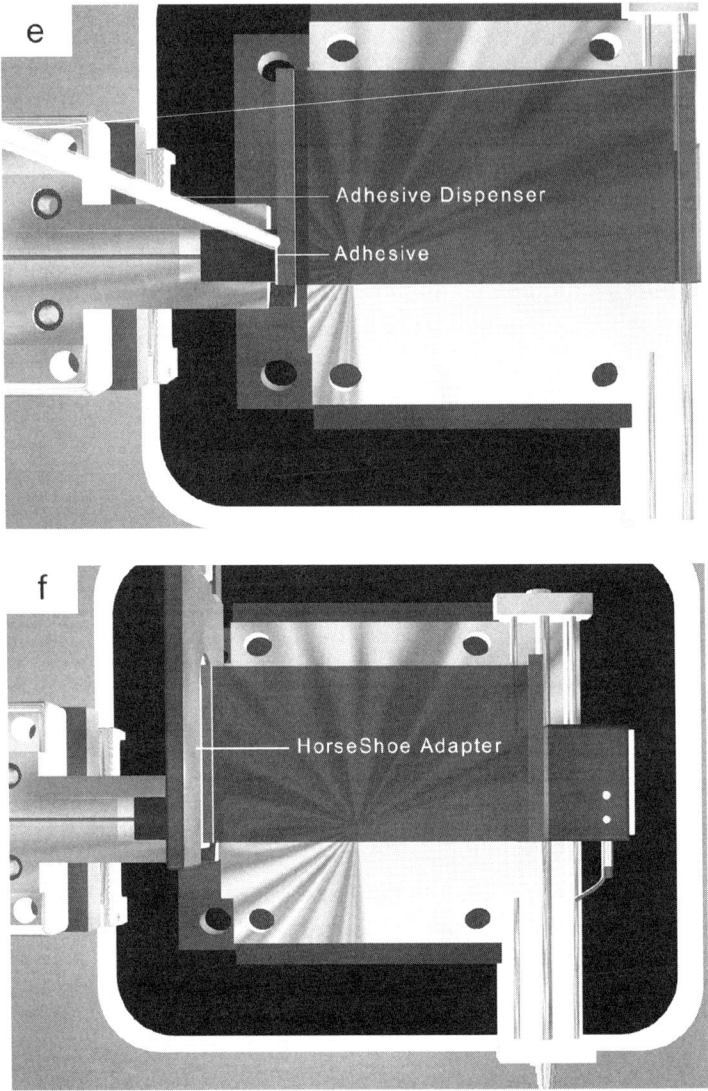

Figure 5.10   (e and f) Step 3 continued

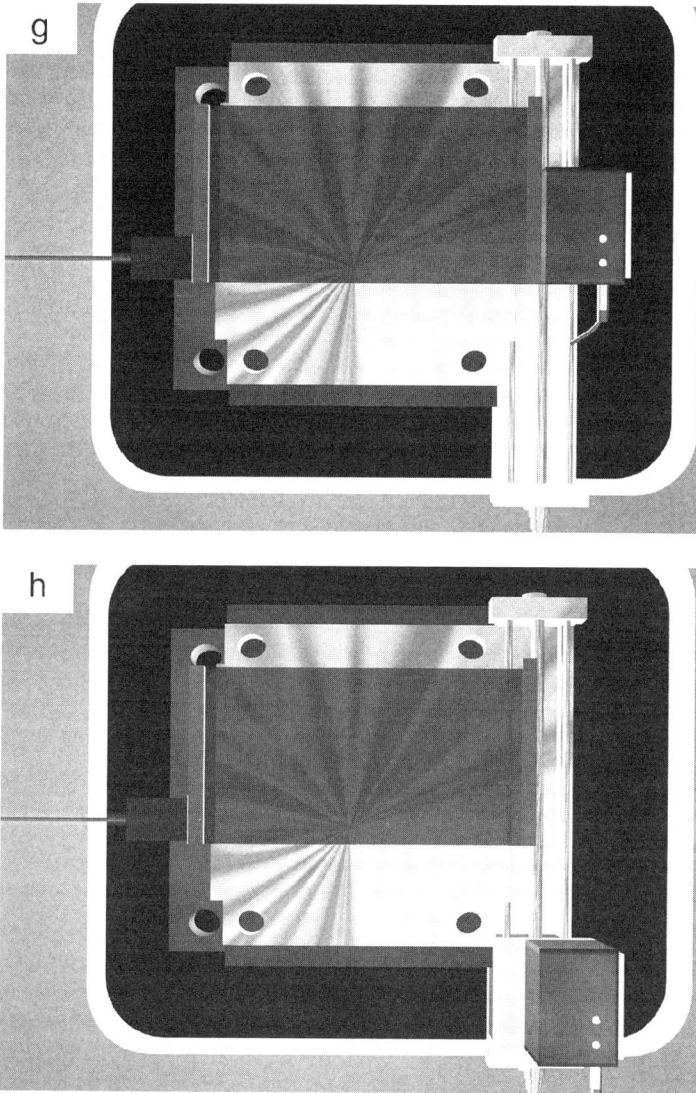

Figure 5.10 (g and h) continued

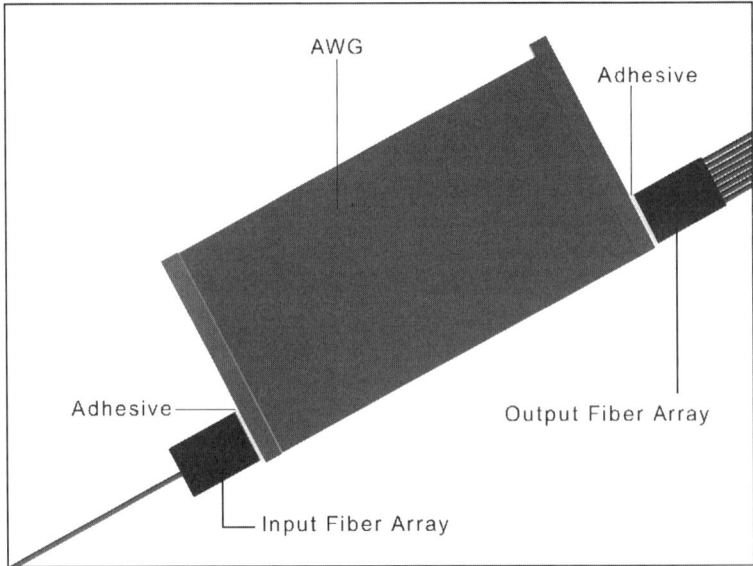

Figure 5.11   PLC attached to IFA and OFA.

Figure 5.9a presents a component assembly workstation for optoelectronic components (as an example, the EXFO ProAlign 5000 is shown). This is a fully integrated workstation for active alignment and assembly of photonics components, especially planar light-guide circuits (PLCs) (e.g., attachment of the input and output fiber arrays to a planar light-guide circuit). The automated light detection system and sequential bonding process in such a workstation minimize insertion loss and improve product yields. Figure 5.9b shows a close-up view of the assembly table. The main constituents (and their functions) of the assembly table include:

(a) Pneumatic stages that are employed for the initial positioning of the input fiber array (IFA) and output fiber array (OFA)

(b) Force sensors that are employed to optimize the spacing between the input and output fiber arrays and the planar light-guide circuit (PLC)

(c) Vacuum chucks that are employed to grip the input fiber array and output fiber array in place during the alignment process

(d) Six-axis nanorobot unit that ensures the PLC alignment

(e) Light detector that is employed for recognizing peak power in the initial alignment of the input fiber array

(f) Translation stage employed to move the light detector into position

The various steps in the assembly process are illustrated in Figure 5.10. Figure 5.10a and b illustrates Step 1: The input fiber array is moved into place by the pneumatic stage. The touch sensor establishes the optimum gap to begin the alignment procedure.

Figure 5.10c and d illustrates Step 2: The six-axis nanorobot unit carries out a search pattern in order to obtain maximum signal. The light detector moves into place to detect light and begin the alignment procedure.

Figure 5.10e and f illustrates Step 3: This step provides bonding of the input fiber array (IFA) to a planar light-guide circuit (PLC). Figure 5.10e shows the dispensing (from the adhesive dispenser, i.e., the syringe in this case) of a UV adhesive on the part. The adhesive wicks in via capillary action to achieve full coverage. Figure 5.10f illustrates cure of the adhesive by employing a custom light delivery system (horseshoe adapter; see Figure 5.8) that delivers UV light.

Figure 5.10g and h shows that, following the completion of cure, the light detector is backed away.

Analogous steps are employed for bonding of the output fiber array (OFA) to a planar light-guide circuit (PLC). These are as follows.

Step 4: In this step, the output fiber array is moved into place by the pneumatic stage, and the touch sensor establishes the optimum gap to begin the alignment procedure. Thus, the IFA–PLC assembly is aligned to the OFA.

Step 5: This step provides the output fiber array (OFA) alignment to a planar light-guide circuit (PLC). The PLC is moved through an alignment routine on the six-axis nanorobot unit across all six axes. Following the realization of a suitable alignment, the nanorobot moves the PLC to the desired position to receive the adhesive bond.

Step 6: In this step, the output fiber array (OFA) is bonded to a planar light-guide circuit (PLC). In this case, the OFA is backed away from the PLC and the adhesive is then dispensed on the OFA only.

Thus, following the procedure outlined above, the PLC is attached to the IFA and OFA (see Figure 5.11). Subsequently, with "pigtailing" completed, the device is ready for the next step of the packaging process.

It should be noted that one of the most important parts in a component assembly workstation is the six-axis nanorobot unit that ensures the PLC alignment. A crucial characteristic of the automated assembly is alignment of the optical components for (i) measurement of power and (ii) final connection. In this context, the most important parameters of the automation system include (see, e.g., Xu et al., 2002):

(a) System stability
(b) Resolution
(c) Bidirectional reproducibility

The system's stability includes the short-term stability on the order of a second for reproducible measurement of power and the long-term stability on the order of 10 min for adhesive bonding and/or device characterization; the system's resolution relates to its sensitivity to controlled changes; and bidirectional reproducibility relates to moving off the initial alignment position and returning to it (see, e.g., Xu et al., 2002). In this context, some of the relevant issues related to the automated assembly process include effects of the following (Xu et al., 2002):

(a) The geometric convolution of the active areas being aligned

(b) Index-matching media in eliminating interferometric power level changes

(c) The materials employed in the complete system and the combined change due to environmental variations

(d) Temperature variations on optical power stability

(e) Repeatability from a displacement back to the optimum power level

(f) Resolution required to hold peak power

Such factors can be evaluated using the setup shown in Figure 5.12 (Xu et al., 2002), which illustrates the measurement of optical power transmission between two fiber arrays, that is, one stationary and one moving that is mounted on an alignment system (FR-3000 NanoRobot). A thermometer with a resolution of 0.1 °C monitors the ambient temperature during the measurements. An example of the measurements performed on such a system is shown in Figure 5.13, demonstrating the optical power dependence on radial displacement.

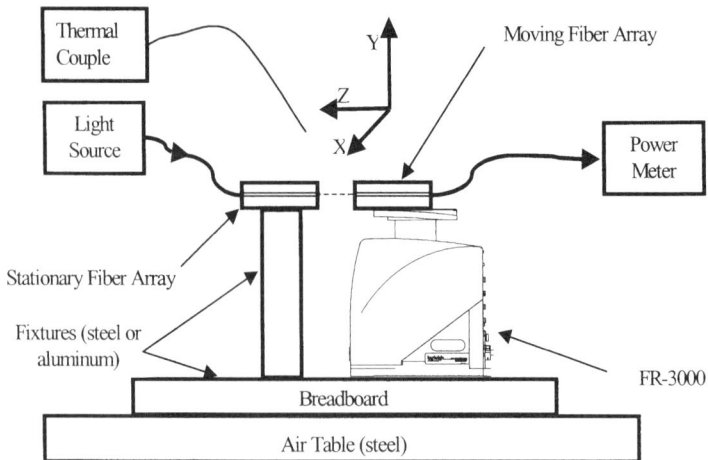

Figure 5.12 Schematic diagram of the setup for measuring optical power through fiber arrays. Reprinted with permission from Xu et al., (2002). © 2002, Burleigh.

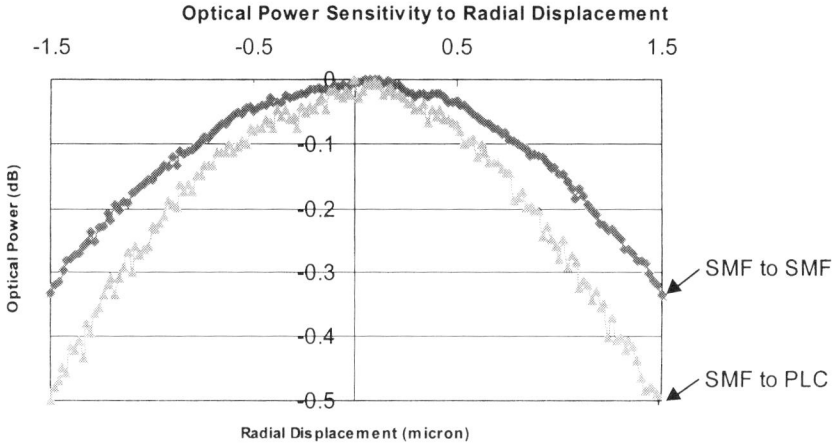

Figure 5.13 Optical power as a function of radial displacement of 20-nm steps (with index-matching fluid added between optical components). Reprinted with permission from Xu et al., (2002). © 2002, Burleigh.

5.4. CURING METHODS AND CONTROL

There are a large number of control variables in the curing process. These include the time–intensity profile of the applied illumination, the spectral range of light used, the number and profile of curing steps, the positioning of the piece (being cured) relative to the light, and the method of distribution of light onto the piece; in addition, all the above should also be related to the characteristics of both the piece being cured and the adhesive. Such interdependence among the various variables of the light-based curing process necessitates its optimization, (i) providing more control over the process and (ii) improving its yield and throughput. Such an optimization for some adhesives may entail, for example, (i) an initial slow curing at low power, (ii) a pause for a given period of time, followed by (iii) curing at a higher power level, which can contribute to greater mechanical stability of the cured joint. In particular, employing such multistep curing profiles for controlling the curing process can help minimize shrinkage and thus help to facilitate alignment (e.g., between fiber-optic components) throughout the curing process. The optimization of the curing process through multistep curing profiles can be accomplished by using a computer program related to a number of separate curing steps, each with its specific cure time, irradiance, and rest interval. Such programmable cure profiles provide a valuable means for optimizing the curing process. The results indicate that employing such programmed cure profiles provides reliable cures in the shortest possible time and with improved adhesive bond properties that meet stringent standards.

For thermally cured adhesives, controlling various temperature profiles during the rapid thermal curing of adhesives is one of the most crucial steps in avoiding thermal runaways. For a curing system and methodology to be integrated in an in-line automated assembly process for high-volume processing in the fiber-optic communications industry, development of systems for automated self-directed curing of adhesives is required to make sure that there is a good bond each time and there is no overcuring or burning.

The self-directed methods developed for curing of various materials in different applications are based on (i) monitoring of information during the curing and repetitive analysis of the temperature in order to maintain a predetermined cure temperature; or (ii) generating and storing a set of cure constants for different compounds in the computer database, so that these can be used later for optimizing the cure of a specific compound; or (iii) comparing actual characteristics to predicted values derived from computer simulations of the cure cycles; or (iv) using the program code for deriving the gradients related to process parameters, which allows for the monitoring and control of processes by determining the rate and direction of change of variables in relation to other process variables. An example of such a self-directed method is illustrated in Figure 5.14.

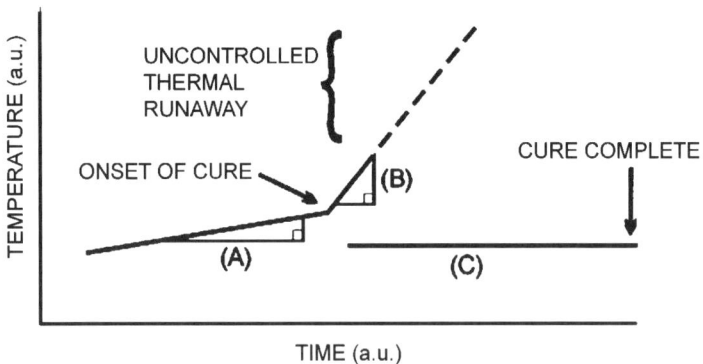

Figure 5.14 Schematic diagram illustrating the self-regulated curing mechanism. Heating (indicated by A) continues until the onset of cure, after which point the exothermic self-heating of the adhesive results in a sharp increase in the heating rate, which may potentially lead to thermal runaway. The key to self-regulation is to determine this onset and reduce the intensity of the cure-initiating irradiation of the adhesive (see region C), thus completing the cure.

5.5. CHARACTERISTICS OF CURED ADHESIVES

5.5.1. KEY CHARACTERISTICS OF CURED ADHESIVES

One of the key characteristics (and potentially most difficult one to determine) of cured adhesives is the precise measurement of the *degree of cure*, which is essential in order to ensure component reproducibility. What is meant by "complete cure," "degree of cure," or simply "cured polymer" may depend on several factors. For example, from the view of an industrial practitioner, "complete cure" may mean the absence of any dimensional change on a macroscopic level or he or she may simply identify fully cured materials as those exhibiting maximum physical, chemical, and thermal properties in use (i.e., in this "*macroscopic perspective*," the cured material is stable from both a dimensional and physical-properties points of view). An analytical chemist, however, may determine the "complete cure" as the absence of any cross-linking reactions in the polymer (i.e., "*molecular-level perspective*"). Thus, one possible criterion for "complete cure" could be some form of stability of a given process, property, or their observation. In this context, other relevant factors could be related to, for example, shrinkage versus creep or environmental factors (e.g., temperature variations) leading to certain effects related to curing (so, does curing as a process on a certain level continue with time throughout the lifetime of a cured assembly?).

In practice, the tests related to hardness and resistance to scratching can be used to establish the degree of cure. In such cases, the hardness value is frequently used as a measure of the degree of cure of an adhesive.

In this context, one of the routinely employed characterization methods is the use of a differential scanning calorimetry (DSC), which provides measurements of heat capacity and the exothermic and endothermic reactions of materials (for more details, see the following section). From these measurements, one can derive the glass transition temperature (T_g) and the degree of cure of adhesives.

5.5.2. CURE ANALYSIS METHODS

Thermal analysis methods are valuable techniques for determining the characteristics of cured materials and optimizing the cure cycle for photopolymerization. In each case, a parameter of the sample is measured as a function of temperature or time while being heated or cooled at a controlled rate. These methods can be divided into three subcategories according to the type of measurement, that is, thermodynamic, mechanical, or electrical.

Thermodynamic measurements are made using *differential scanning calorimetry* (DSC) where the rate of flow of heat energy into a sample is monitored as the temperature is raised or lowered in a controlled manner. Typically, the electrical energy supplied to the heating element of a furnace, containing the sample, is compared to that supplied to an identical furnace used as a reference. This allows for the determination of the heat capacity of the sample, in addition to thermodynamic events such as the glass transition, crystallization, reaction, melting, decomposition, and outgassing. This, however, can be misleading especially if the material has a thermal activator.

Application of a modulated temperature profile to a sample will allow for the separation of the components of the measured heat flow arising from reversible (equilibrium) and nonreversible events (Schawe, 1995). The heat capacity, glass transition, and melting temperature are equilibrium processes, whereas crystallization, reaction, decomposition, and outgassing are nonreversible events. In this example of the DSC technique, a sinusoidal, sawtooth, or step profile is superimposed over the traditional linear temperature program. The oscillation in the heat flow from a reversible event will be in phase with the oscillation in the temperature but shifted in phase for a nonreversible event. A Fourier transform is used to deconvolute the resulting oscillatory heat flow signal into reversible and nonreversible components.

The mechanical properties of a material can be measured using *dynamic mechanical analysis* (DMA) and *thermal mechanical analysis* (TMA). In the former method, the viscoelastic response of a material to a dynamic load at a fixed frequency is measured as a function of temperature or time. In particular, an oscillatory stress with fixed amplitude is applied to a sample while being heated or cooled at a controlled rate. The viscoelastic response of the material is then determined by monitoring the resulting oscillatory deformation to the applied stress using a sensitive strain gauge. Similar to the modulated-DSC method, a phase shift in the oscillatory stress and strain profiles arises from the elastic and viscous (i.e., damping) responses of the material. The values for the viscoelastic properties obtained using this technique include the storage and loss moduli for different modes of deformation, $\tan \delta$, and the complex viscosity of materials. Measurement of the modulus and the viscosity of a material as a function of temperature will allow for the identification of the glass transition (see below).

Thermal mechanical analysis (TMA) involves measurement of dimensional changes in a material as a function of temperature. The thermal expansion coefficient can be derived using this technique, and the temperature dependence of this value gives another method for determination of the glass transition temperature (which occurs where there is an abrupt change in the coefficient of thermal expansion).

Dielectric analysis measures the capacitance and conductance of a material as a function of time, temperature, and frequency. An alternating electric field is applied across a sample placed between a pair of electrodes, and the current passing through the material is measured. The measured current will be shifted in phase from the applied voltage, where the phase shift, θ, is related to the ratio of the loss factor ε'' to the permittivity ε' [$\varepsilon''/\varepsilon' = \tan\delta = \tan(90° - \theta)$ is called the dissipation factor]. The loss factor and permittivity (or dielectric constant) represent the components of the measured current arising from conductance and capacitance, respectively. The values for each component are given as

$$\text{Conductance} = (I/V)\cos\theta$$

$$\text{Capacitance}(\times 2\pi f) = (I/V)\sin\theta$$

where I and V are the measured amplitudes of the oscillating current and voltage, respectively, and f is the applied frequency.

In this context, it should be noted that, in general, there is no single definitive test for determining the degree of cure, and ideally it would be judicious to apply several analysis methods in order to build up complementary information on the curing process.

Thermal analysis methods are used to measure a number of properties of polymeric materials. A key characteristic of a material relevant for applications in the fiber-optics and microelectronics industries is the glass transition temperature (which, in practice, normally represents a temperature range rather than a specific temperature). The mechanical properties of a polymer can vary substantially at temperatures above and below the glass transition. Therefore, the selection of a material for a particular application would involve consideration of the operating-temperature range for the device.

Typically, the glass transition can be identified from the measurement of changes, for instance, in the heat capacity, elastic modulus, coefficient of thermal expansion (CTE), or dielectric constant of the material (see above). The glass transition is not a phase transition like the melting point of a material; instead, it is a kinetic event that depends on the time scale at which a property of the material is observed. At temperature values below the glass transition, the polymer chains are unable to reorganize within the time scale of the experimental measurement. The temperature at which the glass transition occurs may be defined as the temperature at which the time scale for molecular reorganization is equal to that for the measurement method. Therefore, the glass transition temperature will depend on the rate of heating of the sample in each of the thermal analysis methods described above. For the example of measurements using DMA, the glass transition will depend on the rate of the applied oscillatory stress. In addition, the glass transition does not occur suddenly but across a range of temperature values.

Measurements of the glass transition temperature under a standard set of conditions have been correlated with the degree of cure: The highest value measured corresponds to a completely cured material. It should be noted that the glass transition temperature may not be observed in polymers with a very high degree of cross-linking (i.e., fully cured thermosetting materials). In this case, thermal decomposition will occur before molecular reorganization of the polymer chains becomes sufficiently fast. An alternative indirect method for determining the degree of cure is from measurements of the dielectric constant at a fixed temperature.

A brief summary of the various analysis methods for determining the characteristics of cured materials is provided below. These methods are based on the type of measurement, that is, thermodynamic, mechanical, or electrical.

Differential scanning calorimetry (DSC) measures heat flow into (endothermic) or out of (exothermic) a sample as it undergoes a phase change, and heat capacity and T_g are determined.

In dynamic mechanical analysis (DMA), the viscoelastic response of a material to a dynamic load is measured as a function of temperature or time, and T_g is determined. In thermal mechanical analysis (TMA), dimensional changes in a material are measured as a function of temperature, and the coefficient of thermal expansion (CTE) is derived.

In dielectric analysis (DEA), an alternating electric field is applied across a sample placed between a pair of electrodes, and the current passing through the material is measured. Thus, electrical characteristics (capacitance and conductance) as a function of time, temperature, and frequency are derived. The capacitance (i.e., the ability of a material to store an electrical charge) determines the electrical behavior of rigid materials (e.g., a polymer below its T_g). The conductance (i.e., the ability to transfer electrical charge), on the other hand, becomes important in the case of a heated material losing its rigidity (e.g., a polymer above T_g). Thus, such changes in electrical behavior associated with phase changes facilitate an understanding of the structural properties of the material. It should be noted that, typically, the dielectric analysis method has greater sensitivity for detecting small variations, as compared with differential scanning calorimetry or dynamic mechanical thermal analysis.

5.5.3. MEASUREMENT OF SHRINKAGE IN CURED ADHESIVES

A typical application for UV-cured adhesives is in the active alignment of components where the substrates are positioned to a very high accuracy to optimize optical throughput before the adhesive is applied and then cured. However, residual stresses in the cured material can lead to misalignment of the positioned components (see Figures 5.15 and 5.16). The residual stress in thermally cured adhesives will be a function of a number of factors, such as

(i) Cure shrinkage

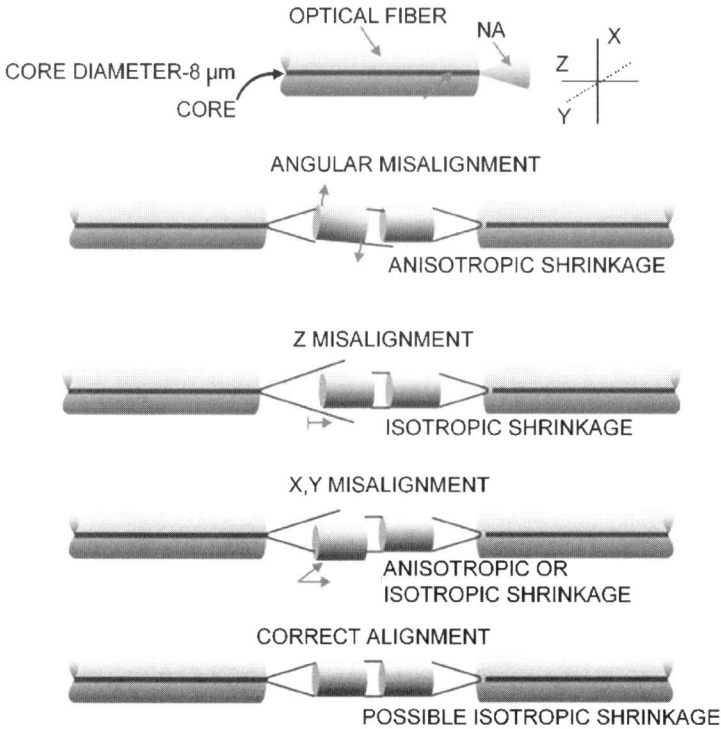

Figure 5.15 Schematic illustration of optical element misalignment due to adhesive shrinkage. Note that the correct alignment can be achieved in systems with repeatable isotropic shrinkage along the optical axis, provided the degree of shrinkage is well characterized.

 (ii) Tensile–elastic modulus
 (iii) Coefficients of linear expansion
 (iv) Uniformity of the cure

An effective method to prevent misalignment during the curing process is to minimize the shrinkage of the material during the polymerization reaction.

Shrinkage is one of the most critical parameters for fiber-optic applications of adhesives. A method for measuring shrinkage in UV-cured adhesives (in the volume of very small quantities of material that are employed in the fiber-optics and microelectronics industries) is based on an optical technique that has been specifically developed to measure small changes (Hudson et al., 2002). It was demonstrated that reproducible values for the volumetric shrinkage could be achieved using spot-curing methods. The use of such techniques will assist in the selection of both the appropriate adhesive for specific applications and the

ALIGNED COMPONENT

ILLUMINATION

ADHESIVE

SUBSTRATE

COMPONENT
TILTS

ILLUMINATION

ADHESIVE
SHRINKS
ANISOTROPICALLY

SUBSTRATE

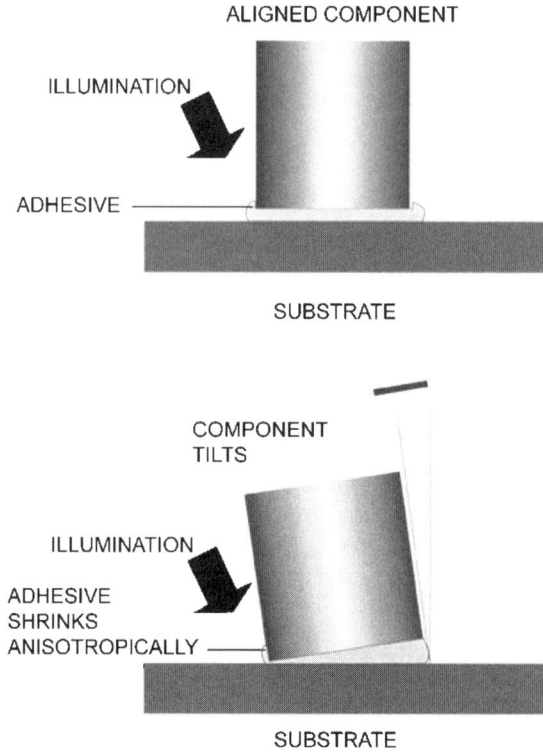

Figure 5.16 Schematic illustration of optical element misalignment due to anisotropic adhesive shrinkage due to shadowing effect causing unequal light intensity to reach different regions of the bond line. This causes anisotropic shrinkage and misalignment.

curing conditions needed to optimize its final properties. The apparatus used in this method (referred to as *drop-shape analysis*) is illustrated in Figure 3.2. A symmetric ("sessile") drop of an uncured thiolene adhesive is formed on a substrate coated with a polymeric material. The adhesive is cured using UV light in the wavelength range between 320 and 500 nm. The radiation is directed onto the sample using a light guide delivering a constant intensity of UV light of 1000 mW/cm^2 at the head of the light guide, and the duration of the exposure is 30 s.

Digital images of the adhesive in a plane perpendicular to the substrate surface are recorded initially and then following exposure to UV radiation. The data for the uncured drop are fitted to a Laplacian profile in order to determine the surface tension of the material. This value is compared with that provided by the manufacturer for the adhesive to give an indication of the accuracy of the image analysis.

It is important to obtain a good contrast between the drop and the surroundings in the recorded images. In addition, the Charge-Coupled Device (CCD) camera and lens need to be oriented slightly out of the horizontal plane in order to identify the points of contact between the material and the substrate surface. If the contrast is poor or the angle for the imaging camera is too large, then the data points collected do not accurately reflect the cross section of the drop, and, consequently, an incorrect value for the surface tension is calculated. An average value of 8.7% for the volume shrinkage of this adhesive is obtained with a fairly small scatter of $\pm 0.2\%$. An advantage of using UV over thermally cured adhesives is apparent in this example. It eliminates the effects of thermal expansion of the substrate, which may lead to considerable residual stress when the system is cooled.

Shrinkage measurements during the curing of adhesives with UV light have been performed for many years on bulk samples of material using dilatometry experiments (Pappas, 1992). These, however, will not accurately describe the volume dispensed for fiber-optics and microelectronics applications. In addition, there are also some practical difficulties related to the need for obtaining accurate measurements from the wet adhesive surface with the probe stylus. To address these difficulties and provide a method to accurately characterize shrinkage of thin films and small drops, which is typical of the amounts of material used in photonics manufacturing, another method has been developed based on a noncontact optical metrology technique, *optical profilometry*. An optical profilometer uses interferometric techniques to accurately measure the surface contour of small features by counting interference fringes. Several laboratory instrument manufacturers make turnkey systems (including computer software) to generate a full three-dimensional view from the surface data. Properly prepared drops of adhesive can thus be mapped before and after cure, providing surface data that may be used to calculate the volume change due to cure shrinkage. Data from this technique correlate very well with the traditional methods and sessile-drop measurements described above.

The adhesive properties that should be considered in various applications in photonics manufacturing include:

(a) Cure shrinkage
(b) Adhesive strength
(c) Operating temperature range
(d) Index of refraction
(e) Viscosity
(f) Elongation
(g) Hardness
(h) Modulus of elasticity

   (i) Tensile strength
   (j) Lap shear strength
   (k) Peel strength
   (l) Cohesive strength
  (m) Outgassing
   (n) Moisture resistance
   (o) Glass transition temperature (T_g)
   (p) Thermal conductivity
   (q) Thermal expansion coefficient
   (r) Rate of creep
   (s) Cure time
   (t) Degree of cure
   (u) Degree of cross-linking
   (v) Shelf life

There are curing profiles that will optimize some but not all of these parameters. It is important (i) to understand which parameters are the most significant for each application, (ii) to choose an appropriate adhesive that generally fits the requirements, and (iii) to optimize its performance with a custom cure profile.

5.6. SPECTROSCOPIC MONITORING OF ADHESIVE CURING

As mentioned above, an important issue, as well as a difficult one to resolve, related to adhesive curing is the determination of the complete cure of the adhesive. In principle, several different methodologies can be applied in order to determine the extent of cure of adhesive materials or polymers. Some *in situ* on-line techniques include, for example, (i) *dielectric cure monitoring*, (ii) *ultrasonic sound speed measurements*, (iii) *infrared absorption spectroscopy*, (iv) *Raman spectroscopy*, (v) *monitoring the optical constants of the film* (i.e., index of refraction *n* and extinction coefficient *k*) using *spectroscopic ellipsometry*, (vi) *fluorescence measurements, and* (vii) *cure-indicating dyes displaying a color change.*

In *dielectric cure monitoring*, variations in the cure state of a polymer adhesive are determined by measuring the dielectric properties of the material being cured. In this method, remote dielectric sensors are used for the measurements conducted in actual processing systems, such as autoclaves and ovens.

Another method is related to measurements of the *ultrasonic sound speed*, which can be correlated to the cure state of the material. This is accomplished by monitoring the velocity of acoustic waves in the material being cured. These techniques are best suited for curing of composites. The method that employs *cure-indicating dyes displaying a color change,* on the other hand is somewhat ambiguous, relying on qualitative observations.

Spectroscopic measurements allow for a more quantitative description of degree of cure by monitoring molecular changes in the material such as the number of double- to single-bond conversions. Recent progress in the availability at reasonable cost of compact spectrometers, narrow-band filters, very low noise and sensitive detectors or detector arrays, and powerful computers has made these techniques more viable and nearly real time.

One of the techniques that is most suited to applications involving optical radiation processing is *fiber-optic Raman spectroscopy*, which is capable of observing the process in real time; thus, curing parameters can be regulated to avoid runaway exotherms or to ensure more uniform curing depth profile. Raman spectroscopy can provide *in situ*, real-time nondestructive monitoring of chemical fingerprint, crystallization, and adhesive curing (see Figure 5.17). This may, in principle, also be achieved with submicrometer spatial resolution. In addition, Raman spectroscopy provides the capability for monitoring various polymer properties, including density and composition, and it can also be used

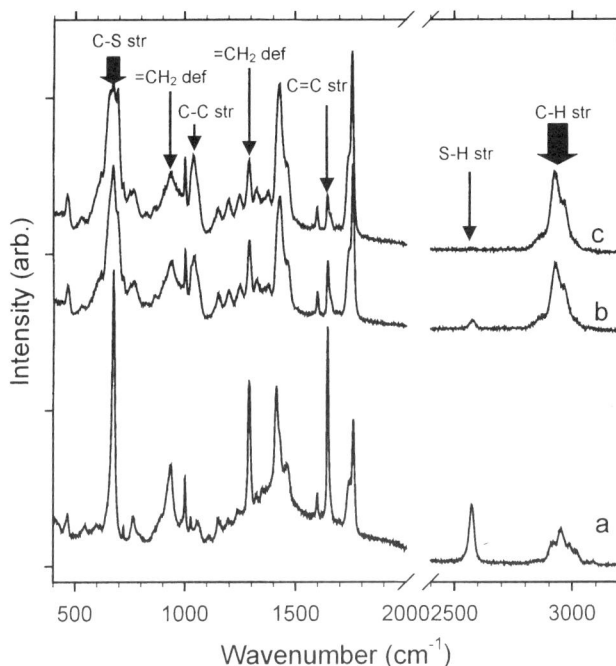

Figure 5.17 Raman spectra of UV-cured thiolene; 5000 mW/cm^2 for (a) 0 s, (b) 0.2 s, and (c) 1.5 s. Raman spectra were collected under 782-nm excitation and normalized to the area under the phenyl peak at 1598 cm^{-1}. In the spectra, "str" stands for stretch and "def" is for deformation. Courtesy of D. V. Heyd and B. Au, Ryerson University, Toronto.

to evaluate the kinetics of curing and possible degradation of the materials. The main advantage of Raman spectroscopy is that the spectra typically provide more precise and characteristic information related to molecular bonding, as compared to IR spectroscopy, but they are also complementary. Furthermore, performing Raman spectroscopy in the IR region using Fourier transform spectrometers reduces interference from fluorescence and provides high sensitivity. In the case of adhesive curing, such a noninvasive monitoring of the material being cured is highly advantageous. In addition, this technique is capable of *remote sensing* using special fiber optics to deliver the excitation to the sample and to transmit the Raman signal to the detector. The ability to monitor and control the curing process is critical in many applications, since the low thermal conductivity and associated internal heat generation in the material being cured can result in temperature and reaction gradients, runaway exotherms, and residual thermal stresses.

Another promising technique for the monitoring of photopolymerization processes is real-time infrared spectroscopy, which allows for the determination of the kinetic parameters, such as the actual rate and quantum yield of the polymerization (Decker, 1990), with millisecond time scales. Such time-resolved IR spectroscopy is especially suited to investigations of the laser-irradiation-induced kinetics of ultrafast chemical modifications in the polymer system (Decker, 1990). For example, from kinetic studies, the photopolymerization of acrylic systems, employing a pulsed or cw ultraviolet laser, was shown to occur almost instantaneously, as recorded by real-time infrared spectroscopy in the millisecond time scale (Decker, 1990). Similar studies also demonstrated the power of time-resolved IR spectroscopy in analyzing the kinetics of UV-induced photopolymerization reactions in polymer systems (Scherzer and Decker, 1999). It was also demonstrated that such UV laser-assisted processing provides an efficient method for realizing rapid (i.e., a few milliseconds) curing of photosensitive resins (Decker, 1999).

5.7. SUMMARY

The types of radiation curing methods include optical radiation, microwave radiation curing, and electron-beam curing. The curing of adhesives using optical radiation is one of the most effective methods for the accelerated curing of various adhesives in applications such as bonding components in microelectronics and fiber-optics for the telecommunications industry. Optical radiation curing is divided among ultraviolet (UV), visible, and infrared (IR) types, corresponding to about 200–400 nm for the UV, 400–750 nm for the visible, and 750–10,000 nm for the IR. A combination of UV and visible light can also provide improved cure rates and depths and facilitate the so-called photothermal processing of materials. Light-based systems are able to deliver faster cures for adhesives than traditional assembly methods. Efficient delivery of energy from a lamp source to the material is achieved by using either a light guide with a narrow diameter or a focusing lens. This technique is referred to as *spot curing*, and, in this case, the light energy is deposited in a localized area where the adhesive has been dispensed.

In some cases, application of a variable-intensity profile is an effective means to optimize the curing process. The delivered intensity of light is varied as a function of time in a series of discrete steps. Such a procedure is referred to as *Step Cure*.

Various analysis methods for determining the characteristics of cured materials are based on the type of measurement, namely, thermodynamic, mechanical, or electrical. One of the key characteristics of cured adhesives is the precise measurement of the degree of cure. One possible criterion for "complete cure" could be some form of stability of a given process, property, or their observation. In practice, the tests related to hardness and resistance to scratching can be used to establish the degree of cure. One of the characterization methods is differential scanning calorimetry (DSC), which provides measurements of heat capacity and the exothermic and endothermic reactions of materials and from which one can derive the glass transition temperature (T_g) and the degree of cure of adhesives. The mechanical properties of a material can be measured using dynamic mechanical analysis (DMA) and thermal mechanical analysis (TMA). Some *in situ* on-line techniques for determining the extent of cure of adhesive materials or polymers include (i) *dielectric cure monitoring,* (ii) *ultrasonic sound speed measurements*, (iii) *infrared absorption spectroscopy*, (iv) *Raman spectroscopy*, (v) *monitoring the optical constants of the film* (i.e., index of refraction n and extinction coefficient k) using *spectroscopic ellipsometry*, (vi) *fluorescence measurements, and* (vii) *cure-indicating dyes displaying a color change.*

One of the important trends in the large-scale production capabilities of the fiber-optics industry is the development of a low-cost automated assembly process. Such systems also have to incorporate in-line means for accelerated curing of adhesives during various stages of the assembly process.

CHAPTER 6

Applications of Adhesive Bonding in Photonics

CONTENTS

6.1. INTRODUCTION

The use of adhesives in optics has a long history from the cementing of doublets for improved optical performance to the mounting of prisms and gratings in spectrometers to the present applications of adhesive bonding in the manufacture of various photonics structures, devices, and assemblies (with demanding performance characteristics), which involve the joining of various materials having different properties.

There are several advantages for using adhesives in photonics applications. These include (i) more uniform distribution of stresses, (ii) ability to adapt to unusual geometries, (iii) providing shock and vibration damping, (iv) low cost, and (v) suitability for high-volume production. There are, however, several limitations to adhesive bonding such as (i) the need for surface preparation, (ii) the need for expensive fixtures and alignment jigs for assembly, (iii) susceptibility of adhesives to various solvents, (iv) limited shelf life and working life, and (v) limited temperature range.

Adhesive Bonding in Photonics Assembly and Packaging
© 2003 by American Scientific Publishers

Although adhesives, especially photocurable adhesives, are still regarded with some reservation in the fiber-optics industry, they are being used in an increasing manner. This is a consequence of the industry's need for high-volume and cost-effective methods of manufacturing fiber-optic components and devices.

This chapter describes the issues related to the selection of adhesives for specific applications, the importance of controlling the curing process in order to achieve a reproducible product, the important parameters for photonics applications, and, finally, some examples of applications in optics, fiber optics, and optomechanics.

Some of the technical issues of concern related to using adhesives in photonics applications are briefly outlined below.

While the microelectronics industry has developed a reliable infrastructure of assembly processes and standards to meet growing demand, fiber-optics manufacturers are still relatively undeveloped in this area.

One should also note some fundamental differences in the use of adhesives for microelectronics and photonics applications. In microelectronics, these concerns are related to the presence of stresses and the issue of structural integrity, as well as the need to maintain high electrical conductivity. In photonics applications (e.g., coupling of two optical fibers), the placement precision considerations are more stringent; such considerations (i.e., alignment issues and the long-term stability of the optical alignment), directed at minimizing any transmission losses of light at the joint, are of paramount importance.

Adhesives, both UV and thermally cured types, afford great opportunity to improve assembly processes in this industry. One of the biggest challenges, however, has been dealing with the issues that remain from the standards that were originally put in place to address the rigorous requirements of the environment for components that are used in submarine and long-haul applications buried under the ocean floor or deep in the ground.

There are several advantages that can be realized by using adhesive bonding, including speed of attachment, uniform distribution of loads, and providing a "seal" against the environment. Adhesive materials, however, can exhibit shortfalls in physical properties for precision-aligned structures, most notably by misaligning the assembly through the shrinkage that accompanies polymerization or by failing to provide a bond in humid conditions. Of particular interest to the fiber-optics industry is the ability of the bonding material to withstand the test conditions of the GR-1209-CORE and GR-1221-CORE testing requirements (administered by Telcordia Technologies). Completed device testing includes the following:

(a) Mechanical shock
(b) Variable-frequency testing
(c) Thermal shock testing
(d) High-temperature storage testing (dry heat—85 °C)

(e) High-temperature storage testing (damp heat—85 °C, 85% relative humidity)

(f) Temperature cycling

(g) Cyclic moisture resistance

(h) ESD testing

Of these, the accelerated tests stress adhesive materials to limit by exposing them to 85% relative humidity (RH) at 85 °C for up to 2000 h.

The adhesives commonly used in fiber-optic applications are thermally cured materials that require heating in an oven, often for protracted periods, to achieve the adhesive bond. Several of these materials are widely used to produce assemblies that pass the stringent testing required by Telcordia (Bellcore). Newer UV materials are becoming more attractive due to the drive toward improved manufacturing techniques. However, despite improvements in these materials, there remain doubts about their long-term performance, in particular, about their ability to pass accelerated testing protocols.

A wide variety of optical, fiber-optic, and optoelectronic components and devices are employed in various applications. Having various sizes and tolerances, these components and devices are assembled into a package (or module) designed to include the coupling of the electromagnetic radiation into and/or out of optical fiber cables. Such optical fibers and components are most commonly joined with adhesives, which provide great flexibility and versatility for joining dissimilar materials. In order not to lose power at the interface, the optical adhesive is selected so that its index of refraction is matched to the index of refraction of the optical fiber. Typically, optical radiation curing of adhesives is the preferred method, providing high cure speeds and throughputs.

One of the critical issues in the manufacture of various optical assemblies is that of automation. In this context, the issues of concern relate to the wide variation in package designs and the lack of standardized alignment and joining methodologies. In addition, assembling optical components generally requires very high precision and placement accuracy (of less than 1 μm). In such cases, automated assembly, involving machine-assisted alignment and attachment processes that facilitate large-volume manufacturing, provides possible solutions to these issues.

It would be instructive at this juncture to analyze a simple fiber-optic application in terms of the alignment sensitivity required. As an example, we will use the packaging of a fiber array to collimator lens array (see Figure 6.1). The sensitivities to misalignment have been analyzed by Zhou and Shi (Lin et al., 2002; Zhou et al., 2002). The various possible misalignments and sensitivities are shown in Table 6.1.

As can be seen from the table, this relatively simple device has very tight tolerance on the roll angles for misalignment. It should be emphasized that after the device is aligned and bonded it must be subjected to the required

Fiber array   Lens array

Top view of the   Side view of the
collimator array   collimator array

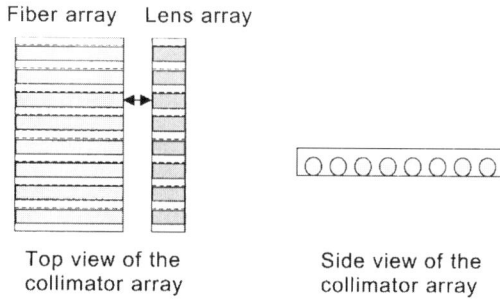

Figure 6.1   Fiber collimator arrays. Reprinted with permission from Zhou et al., *IEEE Trans. Advanced Packaging*, 25 (2002). © 2002, IEEE.

environmental testing as discussed previously and maintain its performance. When the same analysis is performed on the alignment of fiber arrays to arrayed wave guides, the sensitivities for misalignment become more stringent by an order of magnitude due to the mismatch in mode diameters between the fiber and the wave guide. One should compare this with the application of an adhesive

Table 6.1

**Possible Misalignments and Sensitivities of Fiber
Array to Collimator Lens Array**

Loss	1 dB	2 dB
Lateral offset (μm)	2.25	3
Transverse offset (μm)	2.25	3
Longitudinal offset (μm)	6.5	7.5
Pitch offset (°)	11	16
Yaw offset (°)	1.6	2.46
Roll offset (first and ninth fibers) (°)	0.15	0.25
Roll offset (fourth and sixth fibers) (°)	0.6	0.9

used to bond two lenses together to form a doublet, which requires strict control of the index of refraction of the adhesive and its optical transmission but relaxed tolerances for misalignment.

Some other applications of adhesive bonding include fiber bonding, pigtailing, optical component mounting, potting, strain relief, encapsulation, packaging connectors, and fiber-to-silicon bonding. These are employed in photonics products, such as transmitters, amplifiers, receivers, multiplexers, filters, optical add/drops, isolators, wavelength division multiplexing (WDM) couplers, modulators, attenuators, and fiber Bragg gratings.

6.2. ADHESIVES, METHODS, AND APPLICATIONS

With the drive by fiber-optics manufacturers to increase throughputs and yields, the obvious step is to review areas in the manufacturing process that are both time consuming and labor intensive. In general, the component assembly process is very much a manual process, involving fine placement and alignment of active and passive components before the dispensing of an adhesive and the curing stage. The fixturing of these components is of great importance, as alignment must be maintained throughout the curing process (see Figure 6.2).

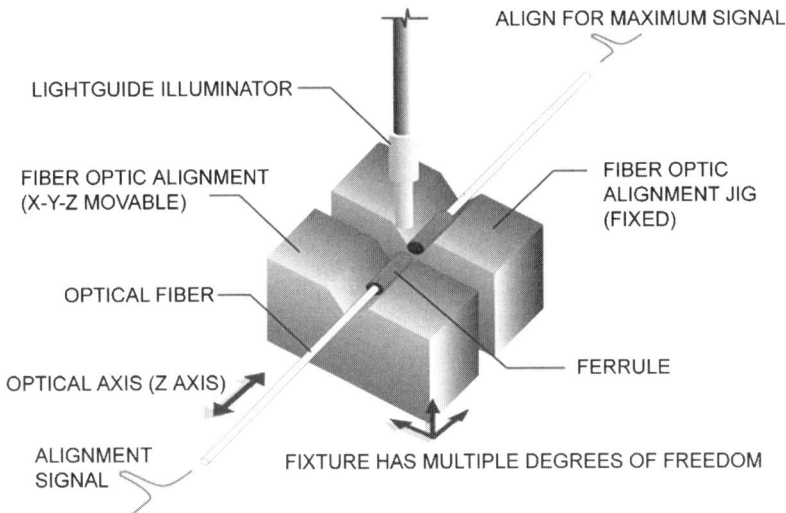

Figure 6.2 Schematic illustration of precision active alignment, showing multiple-axis stages, input and output signals, and light guide employed to cure adhesive. Up to six axes must be monitored before the adhesive is dispensed and cured. UV adhesives are commonly used due to their high curing rate, control, and ease of use.

As mentioned above, many of the devices in the industry use thermally cured materials and are also assembled in batches. The problem with this process is that if the fixturing does not maintain the alignment, or the adhesive shrinks or creeps during the curing process, the assemblies will become misaligned with the obvious possibility for an entire batch of parts to fail. In dealing with thermally cured materials, several manufacturers are evaluating infrared spot and microwave localized technologies to improve this process. The advantage of the infrared option is that it has the ability to be performed in line, versus the microwave process, which is performed in batches, similar to oven-curing processes.

The infrared curing method is relatively novel in that energy delivery between 0.7 and 4.5 μm initiates the curing process directly in the adhesive material. This selective curing actually inhibits many of the factors that detract from the batch oven-curing process. As only the adhesive experiences the delivered energy, there is less collateral heating of either the component or the fixture. The other significant benefit to infrared curing is related to the fact that since only the adhesive is being "heated," the cure speed is increased drastically. Results for curing on the order of 10 to 30 s are regularly achieved from original baselines that were established at 30 to 45 min in a convection oven. This process affords much greater control over both the process and the materials.

In the case of UV adhesives, great advances have been made since those early days of component validation when only thermally cured materials were considered worthy of being selected. Mature processes have been developed by medical device and disk drive manufacturers. In the medical device industry, the main targets for the process are control and repeatability of both the adhesive and the curing system. Rigorous testing is applied to the adhesive to be qualified for United States Pharmacopeia (USP) Class VI status and the Food and Drug Administration (FDA) must validate the entire process. Similar considerations have been applied to the fiber-optic component manufacturer as well. In the disk drive industry, those features of control and repeatability are equally as important.

Although the number and type of devices that are being manufactured in the fiber-optics industry are too numerous to list, the adhesive applications and the types of joint designs for bonded assemblies are far more limited. The following are the most common applications of adhesive bonding in the fiber-optics industry:

(a) *Fiber bonding.* A fiber may be bonded to a silicon surface such as a wave guide, to a silicon wafer, or to any other substrate (see Figure 6.3).

(b) *Optical component mount.* Components can be mounted to a base surface such as inside a butterfly package in a laser diode (see Figure 2.12).

(c) *Lens bonding.* A graded-index (GRIN) lens or ball lens can be bonded in an active component (see Figure 6.4).

SILICON WAFER

FIBER

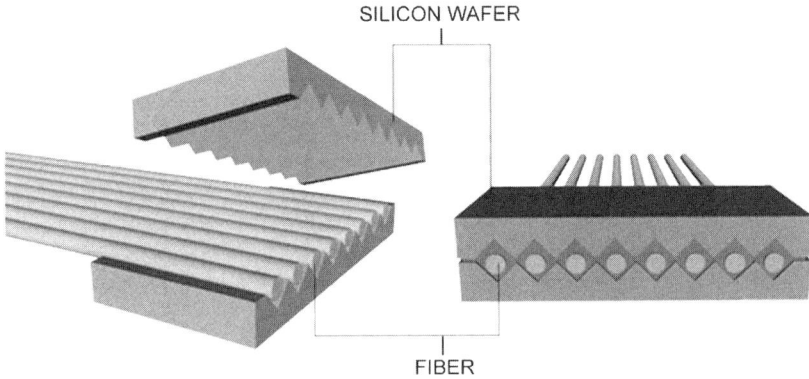

Figure 6.3 Schematic illustration of a precision fiber array. In this case, multiple fibers (i.e., eight) are placed into V-grooves on a silicon substrate, and the adhesive is dispensed into the V-grooves before the top half of the silicon wafer is attached.

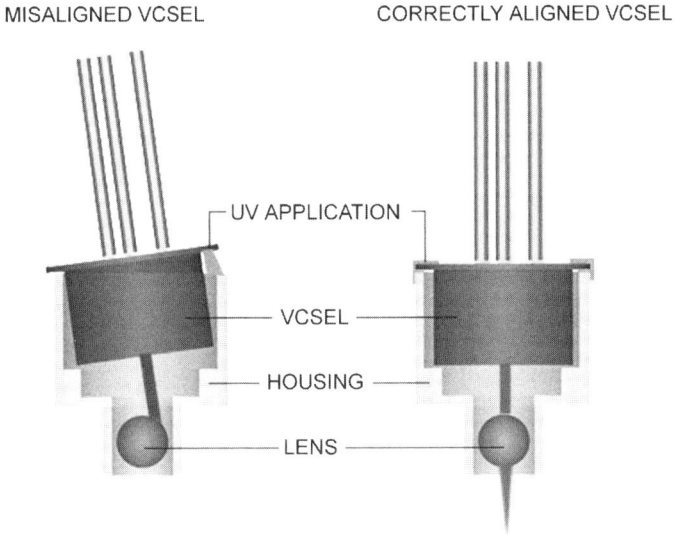

MISALIGNED VCSEL   CORRECTLY ALIGNED VCSEL

UV APPLICATION

VCSEL

HOUSING

LENS

Figure 6.4 Schematic illustration of a vertical-cavity surface-emitting laser (VCSEL), which demonstrates the importance of controlled UV curing. If the adhesive shrinks during cure, the device becomes misaligned (left). A correctly aligned VCSEL (right) can be achieved with an optimized cure profile.

(d) *Encapsulation.* A fiber may be encapsulated in the top of the device such as a vertical-cavity surface-emitting laser (VCSEL) (see Figure 6.4).

(e) *Ferrule attach.* Fiber bonding into a type of connector is one of the most common applications (see Figure 6.5).

Adhesive selection should be based on the substrates to be bonded, the geometry of the joint design, and the environment that the device will experience during assembly and during its final end use. The physical stress that the device will experience will also impact the adhesive choice. Any environmental testing, such as the Telcordia/Bellcore standards, need to be known at this time to make an informed selection.

The critical properties of adhesives used for photonics applications include:

(a) Refractive index
(b) Glass transition temperature
(c) Coefficient of thermal expansion
(d) Absorption spectrum
(e) Outgassing
(f) Percentage shrinkage

For completeness, these properties are summarized below.

The index of refraction is the ratio of the speed of light in a vacuum to the speed in the material and can be measured with a refractometer. The index of refraction of most adhesives is in the range between about 1.45 and 1.6. Typically, in order to minimize Fresnel reflection loss, the adhesive with a refractive index equal to that of the substrate (or very close to it) is used.

The glass transition temperature represents the transition region where the glass transition occurs. T_g represents the transition temperature between the glassy and viscoelastic states of polymers. In general, it represents a temperature

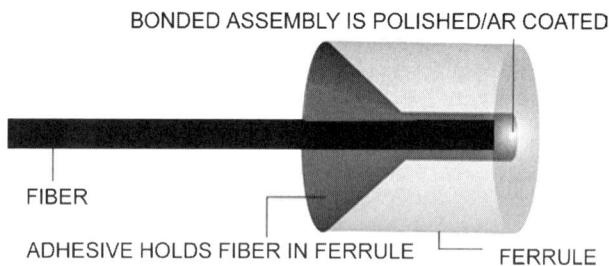

Figure 6.5 Schematic illustration of a typical fiber-to-ferrule attach. In this case, the fiber is inserted into the ferrule, then adhesive is allowed to wick into the assembly via capillary action. The assembly is then cured using conductive, UV, or IR curing, depending on the adhesive selected. Following assembly, the exposed fiber is polished and coated with an antireflective coating.

range. There are several methods that can be employed to measure the glass transition temperature and these include differential scanning calorimetry (DSC), dynamic mechanical analysis (DMA), and thermal mechanical analysis (TMA), and each method provides a slightly different number. In DSC, the rate of flow of heat energy into a sample is monitored as the temperature is raised in a controlled manner. In DMA, the viscoelastic response of a material to a dynamic load at a fixed frequency is measured as a function of temperature or time. TMA involves measurement of dimensional changes in a material as a function of temperature; the coefficient of thermal expansion can be derived with this method and the temperature dependence of this value provides another measure of the glass transition temperature. The relevance of the glass transition temperature to fiber-optic applications involves selecting adhesive materials with a T_g above the normal maximum operating temperature to minimize strength loss during temperature cycling. If the application requires a bond with flexibility over a broad range, then an adhesive with a low T_g should be chosen.

The coefficient of thermal expansion (CTE) is the change in length per unit length of the material for a change of 1 °C in temperature. The CTE can be measured using the technique of TMA. Typically, adhesives with a CTE matching the substrate are chosen to minimize stress on the joints.

The absorption spectrum is measured with a UV/visible spectrometer or Fourier transform infrared (FTIR) spectrometer (in the IR range). Adhesives used in the transmission path in fiber-optic applications should have high transmission at critical wavelengths in the range between 1400 and 1600 nm, corresponding to telecommunications transmitting wavelengths. Also, because of the high power densities used with new pump lasers, the transmission at the second or higher harmonics should also be considered, since 700-nm and 450-nm two-photon absorption or multiphoton absorption could cause adhesive failure.

Outgassing of adhesives is the release of solvents. This occurs during the cure but also throughout the life of any material. The National Aeronautics and Space Administration (NASA) has tested thousands of materials and publishes a Web site with this information (http//epims.gsfc.nasa.gov/og). The two measures of performance are the TML (total mass loss, in percent) and the CVCM (collected volatile condensable material, in percent). NASA has determined that space-qualified materials should have a TML of less than 1% and a CVCM of less than 0.1%. Low outgassing of volatiles is very important for hermetic packages containing active components such as laser diodes as these materials may condense on the laser facets, causing premature failure of the devices.

Shrinkage is a measurement of the change in physical dimension of a polymer from the liquid (uncured) to the solid (cured) state. There are several methods for measuring linear or volumetric shrinkage. These include measuring density, imaging of sessile drops, and interferometric measurements. Typical values for light-cured acrylics are between 3 and 5% and between 1 and 2% for epoxies. These values can be modified using a programmed curing profile known as Step Cure rather than a simple single exposure cure (Martin and Hubert, 2000). Due to the stringent alignment requirements for fiber-optic components, adhesives with low shrinkage values should be chosen.

There is a wide range of both single- and two-component adhesives (e.g., polyester, epoxy, and urethane based) designed for bonding various optical components and assemblies. The appropriate selection of an adhesive for a specific application should always consider component configurations and physical properties (e.g., geometry, materials type, bond line configuration, coefficients of thermal expansion) and optical properties of bonded components (e.g., refractive indices and transmission characteristics of the components). In addition, it is also essential to anticipate any environmental conditions that may be experienced by the bonded optics or photonics assembly (e.g., working temperature limits and chemical resistance and mechanical shock requirement). These demands are also exacerbated by the fact that many of the devices and components being joined are extremely small in size and often are assembled by high-speed automated systems to precise alignment requirements.

To summarize, selection of an adhesive involves evaluating a wide range of variables to meet the demands of a specific application. In fiber-optic applications, these demands are exacerbated by the fact that many of the devices and components being joined are relatively small in size. The major factors that must be considered in selecting an adhesive include:

(a) The assembly of the device or components
(b) The dispensing of the material
(c) The final operating environment of the assembly
(d) The substrates being joined, or encapsulated
(e) The geometry of the area being joined
(f) How the adhesive will be applied
(g) How the material will be cured

It should be noted that, for space, military, and other harsh environments, more stringent requirements on adhesives apply (Lee, 1988). In harsh environments, the main causes of a material's degradation typically include exposure to, for example, temperature, radiation, water (moisture), vibration, and corrosion, which have much greater ranges (extremes) than those expected for typical commercial applications, but generally also apply to the manufacture of fiber-optic devices employed in telecommunications applications.

6.3. ADHESIVE BONDING
IN FIBER-OPTIC TECHNOLOGY

Some examples of the adhesive bonding in fiber-optic technology are outlined below.

A typical case is that of fiber-to-glass-ferrule attachment. In this application, commonly referred to as fiber pigtailing, an optical fiber stripped of its buffer is cleaned and then loaded and secured (bonded) into a ferrule. Ferrules may be metal, ceramic, or glass; glass ferrules are widely used throughout the fiber-optic component assembly industry because they readily adapt to UV curing processes as well as to traditional heat curing. The adhesive is dispensed into the relief at the back end of the ferrule and carried into the narrow tube of the ferrule by capillary action. The adhesive is cured, providing the dual benefits of fixing the fiber into the ferrule and relieving strain at the back end of the ferrule (see Figure 6.5). The front surface of the fiber–ferrule assembly is then polished and may be coated with an antireflective coating. The ferrule is used to improve the handling of the fiber by providing a larger, easily manipulated feature that can be integrated into a component assembly. A typical application for this type of assembly is found in the manufacture of optical add–drop modules (OADMs) such as that illustrated in Figure 6.6.

Typically, the adhesive is an epoxy and may be either UV or thermally cured, depending on whether the ferrule is made of glass or metal.

In another application, namely, port device manufacture, individual fibers (especially single-mode fibers) are difficult to handle and align in an assembly process and thus the fibers are typically secured within a ferrule. Once the fiber is assembled into the glass ferrule, subsequent operations can be used to attach additional components to build up device modules, such as an optical add–drop

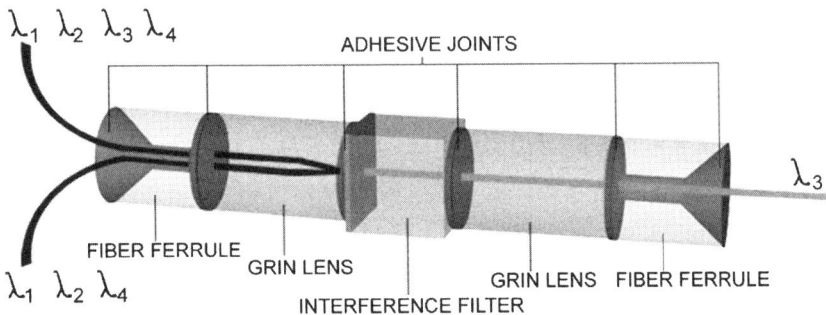

Figure 6.6 Schematic illustration of a typical add/drop module containing several adhesive joints (in this case, the maintenance of alignment is important during assembly). If an adhesive is in the optical path, the refractive index is an important part of the adhesive selection criteria.

module (OADM). In this device, thin-film filters are used to separate discrete wavelengths of light used in dense wavelength division multiplex (DWDM) components, allowing a channel (wavelength) to be dropped or added to the transmission fiber. Thin-film filters (TFFs), which are coated with many ultrathin layers of dielectric material using the technique of ion-assisted physical vapor deposition, are designed so as to reflect or transmit particular wavelengths.

In assembling a three-port OADM, several adhesive joints are used to bond individual components, forming a long line with fiber pigtails at either end. Thus, pigtailing the fibers into capillaries is the first step, as outlined in the first case above. Once the fibers have been secured into the ferrules, collimating graded-index lenses (GRIN lenses) are attached to the ferrules by bonding directly to the front surface of the ferrule. This step requires active alignment where a signal is launched into the fiber and the output measured as the GRIN is aligned to the front of the ferrule, with the objective to fix the lens in position when the maximum throughput is observed. This type of active alignment lends itself very well to UV adhesive applications because of the advantage afforded by *in situ* curing. When the desired alignment has been achieved, the operator can activate the UV curing system and cure the adhesive, firmly fixing the components together. To ensure that the bond does not move during the curing cycle, appropriate delivery optics should be used to evenly distribute the light around the circumference of the bond. In the illustration (see Figure 6.7),

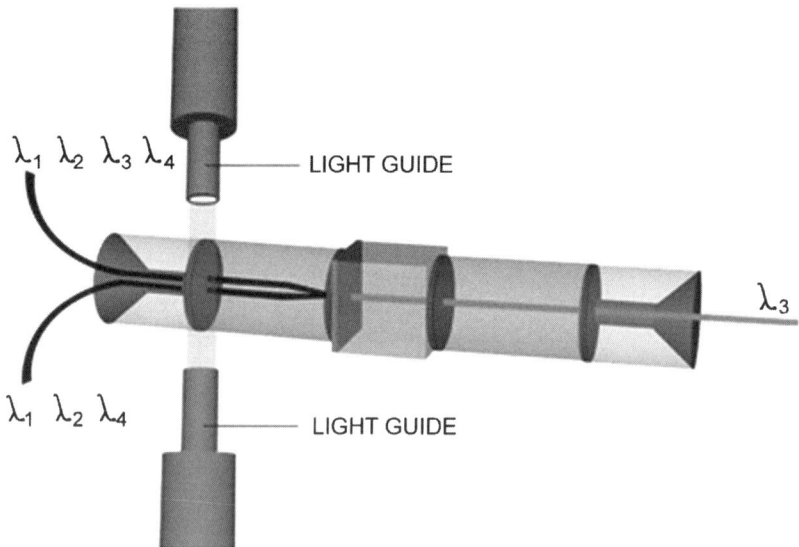

Figure 6.7 Schematic illustration of an add/drop module demonstrating the importance of light delivery during cure. In this case, light is delivered from top and bottom in order to ensure a uniform adhesive cure.

two (or three) light guides are used on opposite sides of the assembly to avoid shadowing effects that may lead to differential curing on one or the other side, an effect that may lead to misalignment (through tilt); alternatively, a cure ring, which provides continuous light around the circumference, could be used. The bonding of the TFF is similar to that used to align the GRIN lens; in this case, active feedback control using a real-time signal is used to ensure maximum throughput. Again, UV adhesive is the optimum choice for this type of application, for all the same reasons as with a GRIN lens. This should be coupled with the optimal Step Cure profile to minimize any material shrinkage. Figure 6.6 shows the assembled device. The completed TFF WDM module is then potted or hermetically sealed and further encased to protect the inner components. In a standard manufacturing process, TFF WDM modules or three-port devices work together in 4, 8, 16, or 32 channels, and they can be packaged together instead of individually. The actual size of a completely assembled TFF WDM module is approximately 30 mm by 5 mm.

As mentioned above, in the case of an optical component mount, various components can be mounted to a base surface such as inside a butterfly package in a laser diode (see Figure 2.12). The optical components are typically attached with suitable adhesives. Such optoelectronic devices are commonly packaged in a hermetic butterfly housing made of low-thermal-expansion materials such as Kovar or a ceramic. Due to the low thermal conductivity of Kovar, heat sinks are typically brazed to the base of the package, which may also contain the photodiode monitor, thermoelectric cooler, thermistor, ball lens, and optical isolator. Some of the important requirements in such a case include stable optical coupling, effective thermal pathway, no outgassing, passive alignment (preferable), and hermetic enclosure (the lid is typically glass or ceramic).

Another common application is that of bonding fiber arrays to collimator lens arrays (Lin et al., 2002; Zhou et al., 2002). Currently, most fiber applications are not limited to bonding single fibers but rather to a one- or two-dimensional array called a fiber array (FA). Care must be taken during the assembly of the array to maintain accurate spacing of the fibers. In an ADM such as discussed above, the light from the fiber must be collimated before passing through the narrow-band filter to prevent broadening of the transmission profile due to angular effects. Therefore, the FA must be aligned and bonded to a collimator lens array (LA) as shown in Figure 6.8. The FA and collimator lens array must be accurately aligned (see Figure 6.9). Misalignment can occur due to lateral offsets or angular offsets. Table 6.1 summarizes the misalignment tolerance of the system. As can be seen from the data, the most critical alignment is the roll alignment (see data in Figure 6.10). Typically, the two arrays are initially aligned using a vision system based on alignment of the edges and optical axes of the arrays. Then, using a six-axis fine-positioning system such as the Burleigh FR3000, the final alignment is performed while monitoring the throughput of each channel.

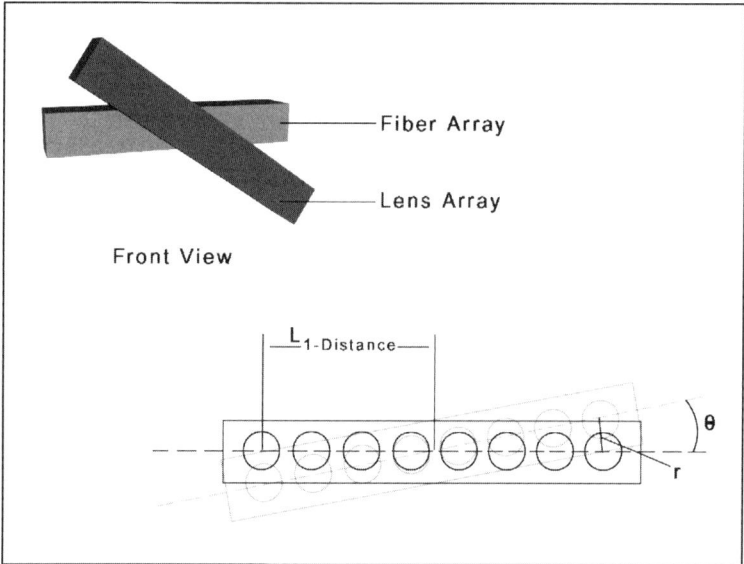

Figure 6.8   Schematic illustration of bonding a fiber array to a lens array. Reprinted with permission from Zhou et al., *IEEE Trans. Advanced Packaging*, 25 (2002). © 2002, IEEE.

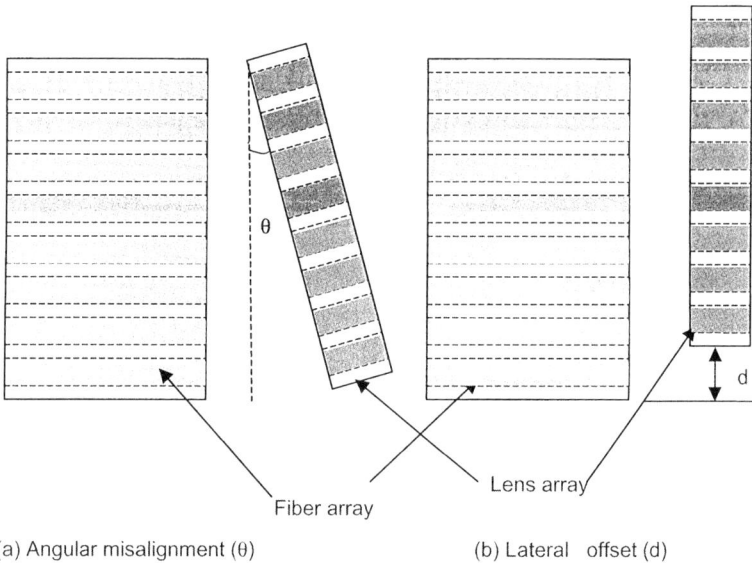

(a) Angular misalignment (θ)   (b) Lateral   offset (d)

Figure 6.9   Schematic illustration of (a) angular misalignment and (b) lateral offset. Reprinted with permission from Zhou et al., *IEEE Trans. Advanced Packaging*, 25 (2002). © 2002, IEEE.

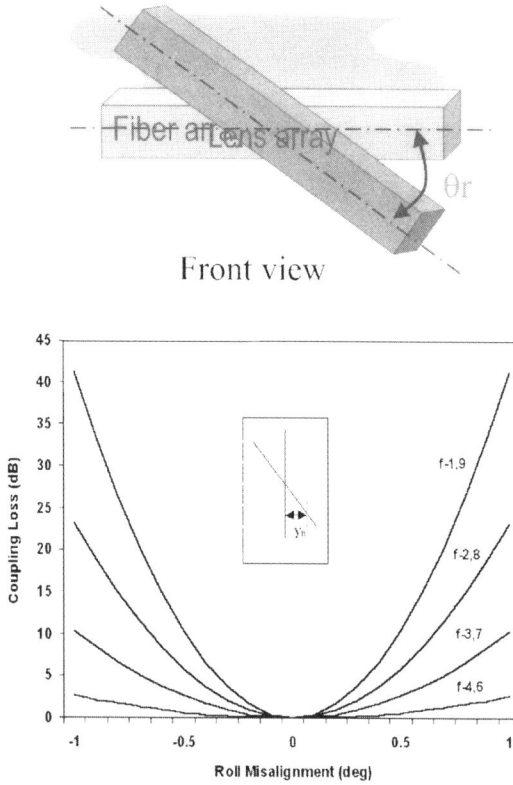

Figure 6.10 Sensitivity to roll misalignment. Reprinted with permission from Zhou et al., *IEEE Trans. Advanced Packaging*, 25 (2002). © 2002, IEEE.

Next, the adhesive is dispensed on the bond line and cured using a UV light source (see the process flowchart in Figure 6.11). During the course of this work, it was found that the bond was more stable and reproducible if a glass plate was attached to hold the FA aligned to the LA as shown in Figure 6.12. For these applications, three adhesives were evaluated, which were all cured with UV light using a dual-step cure, that is, 1000 mW/cm^2 for 1 s, followed by 1500 mW/cm^2 for 20 s. As can be seen from Figure 6.13, adhesive B showed very little change in alignment due to the curing. These authors have identified a method with high precision and high yield using a UV step cure process. The bond can now be subjected to high-temperature storage and 85/85% testing (i.e., damp heat at 85 °C and 85% RH) to determine how stable this bond is under these more stringent environmental conditions.

Assembly Flow Chart

Load 1x8 fiber array (FA) and
lens array (LA)

Align FA and LA to optimal
coupling position

Measure the coupling loss of
each channel

Attach a plate between FA and
LA with adhesive

UV cure the adhesive over the plate to
bond FA and LA together

Measure coupling loss of
each channel after UV cure

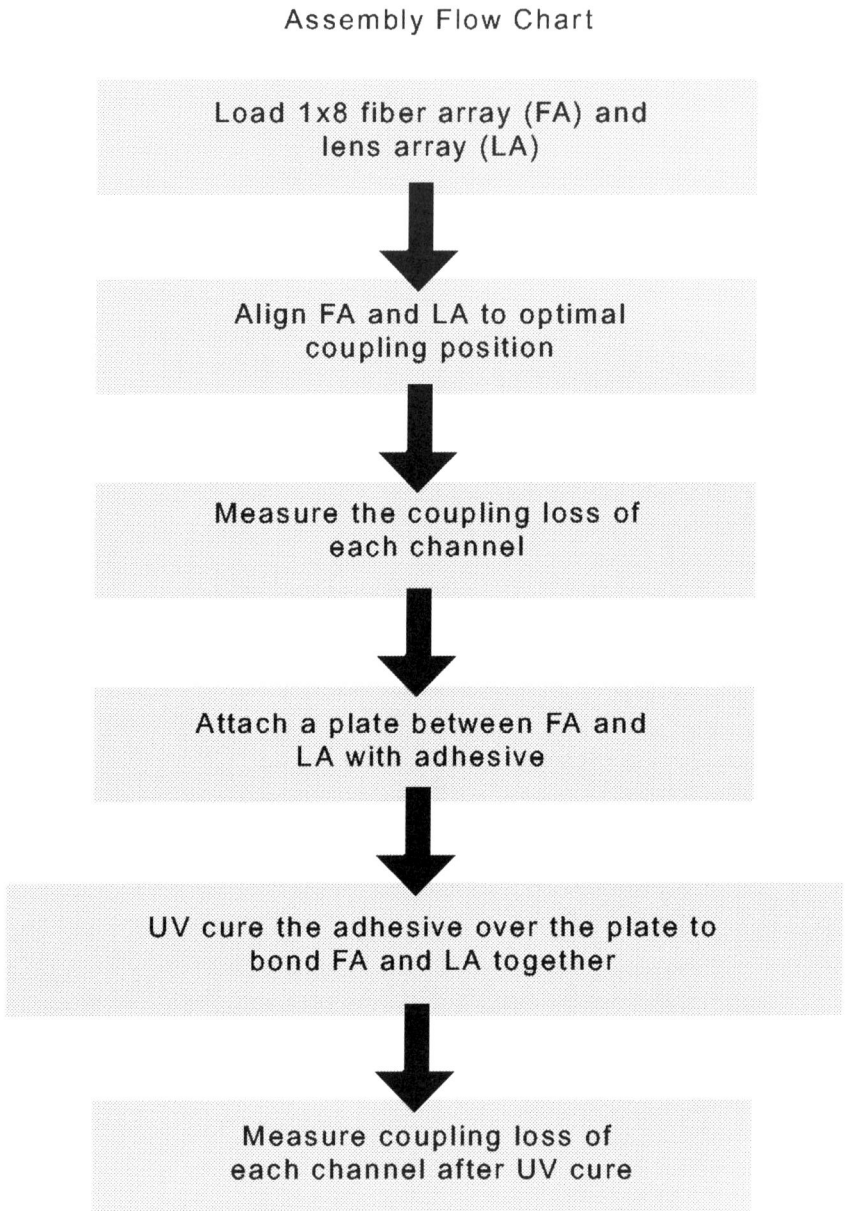

Figure 6.11 Alignment process cycle. Reprinted with permission from Zhou et al., *IEEE Trans. Advanced Packaging*, 25 (2002). © 2002, IEEE.

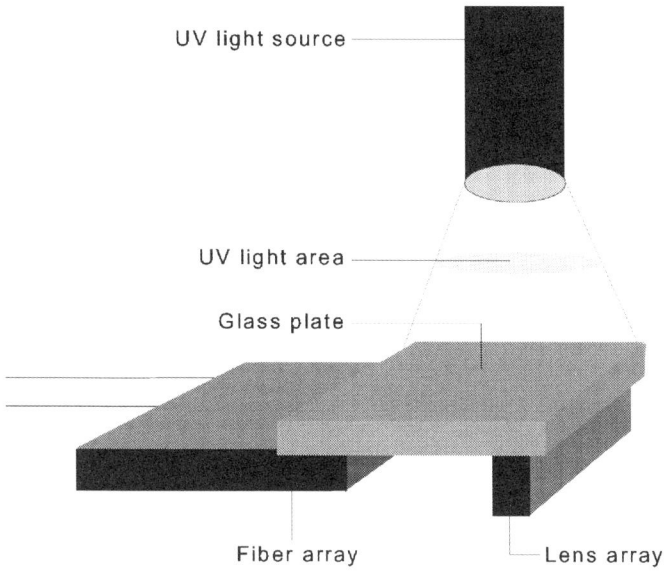

Figure 6.12 Schematic illustration of completed assembly. Reprinted with permission from Zhou et al., *IEEE Trans. Advanced Packaging*, 25 (2002). © 2002, IEEE.

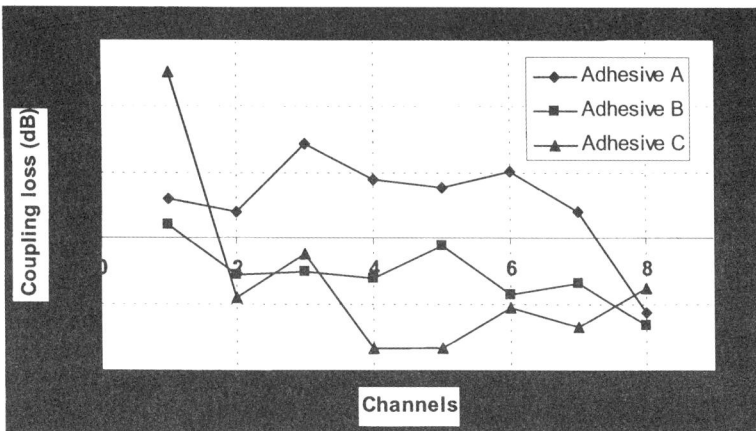

Figure 6.13 Coupling variation due to bonding process. Reprinted with permission from Zhou et al., *IEEE Trans. Advanced Packaging*, 25 (2002). © 2002, IEEE.

Another more complex application is that of bonding a fiber array to an arrayed wave-guide grating. (This study is based on the work of Ehlers et al., 2000.) Wavelength division multiplexing (WDM) systems are well suited to deliver the large bandwidths required for modern telecommunications systems. There are several types of multiplexers that have been developed, including interference filters, Bragg fiber gratings, and planar wave-guide circuits known as arrayed wave-guide gratings. The planar fabrication process allows the realization of high-performance filters with a large number of wavelength channels. In addition, integration with other optical elements such as erbium-doped amplifiers on the same chip seems possible. Arrayed wave-guide gratings (AWGs) are a special kind of planar light-guide circuit (PLC), which are very attractive for DWDM systems because of their great flexibility in filter design. Basically, an AWG is an optical spectrograph built by a planar wave-guide technique that works in a high grating order (50–250) (see Figure 6.14). AWGs with 8, 16, 32, and 64 channels are available, and the major difficulty occurs in aligning the AWG to the matching fiber array. There is a significant mismatch in the mode size for a single-mode fiber, typically 4.9 μm, and that for a single-mode wave guide, typically 3.6 μm (Si/SiO$_2$ core index 1.54 and difference in n between core and clad of 0.7%, 6 × 6 μm core size). Thus, great care is required during the coupling of light from the fiber into the wave guide. General alignment requirements include angular misfits of less than 1°, lateral offsets of less than 1 μm, and longitudinal gaps of less than 10 μm (bond thickness). The structure of the AWG module includes the AWG chip, a pair of fiber arrays, and a chip carrier to control the temperature of the device. The AWG was packaged as follows: the AWG chip was fixed to a heat spreader by using a stress-free mount technique, which employs an adhesive with a low Young's modulus. The heat sink was mounted on a Peltier element, which was attached to the case. The fiber arrays were adjusted and bonded together with a UV-curable adhesive (Panacol Vitralit 7104, but other adhesives can be used). To optimize the coupling of the fiber array to the waveguide array, the chip module and the submount were fixed onto a three-axis stage (x, y, z). The fiber array was also fixed to a stage with 5 degrees of freedom (3 linear, 2 angular). Light from a tunable source was fed via a 1 × N optical splitter to all fibers of the array. The initial prealignment was made manually by adjusting the fiber array in line with the wave guides using a microscope. The output of the opposite side of the AWG was monitored via a microscope lens, split field optic, and infrared camera. The optical outputs near the center of the AWG were aligned. The exterior outputs of the AWG were then aligned more precisely (see Figure 6.15). The laser had to be tuned to the center wavelength of the filter of the observed fiber port. For optimal alignment, two laterally adjustable IR detectors were used. Finally, the adhesive was inserted into the gap between the array and the AWG chip and hardened with UV light. To further stabilize

Waveguide Grating ($\triangle l = m \cdot \lambda_c$)

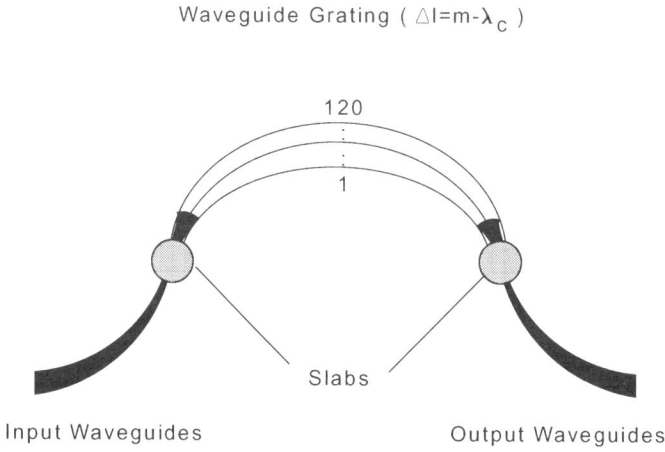

Figure 6.14 Schematic illustration of AWG chip outline. Reprinted with permission from Ehlers et al., *Optical Fiber Tech.* 6, 344 (2000). © 2000, Academic Press.

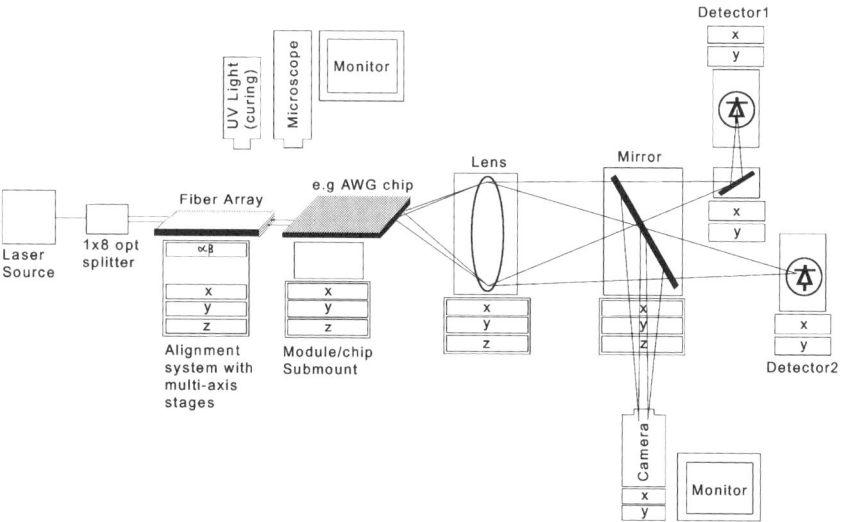

Figure 6.15 Schematic illustration of experimental setup of fiber array–chip coupling. Reprinted with permission from Ehlers et al., *Optical Fiber Tech.* 6, 344 (2000). © 2000, Academic Press.

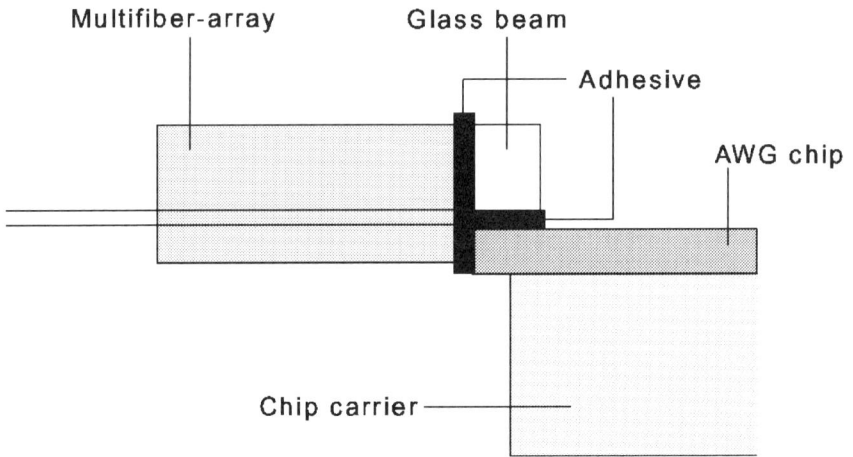

Multifiber-array   Glass beam

Adhesive

AWG chip

Chip carrier

Figure 6.16 Schematic illustration of a sectional view of the fiber array–chip connection. Reprinted with permission from Ehlers et al., *Optical Fiber Tech. 6*, 344 (2000). © 2000, Academic Press.

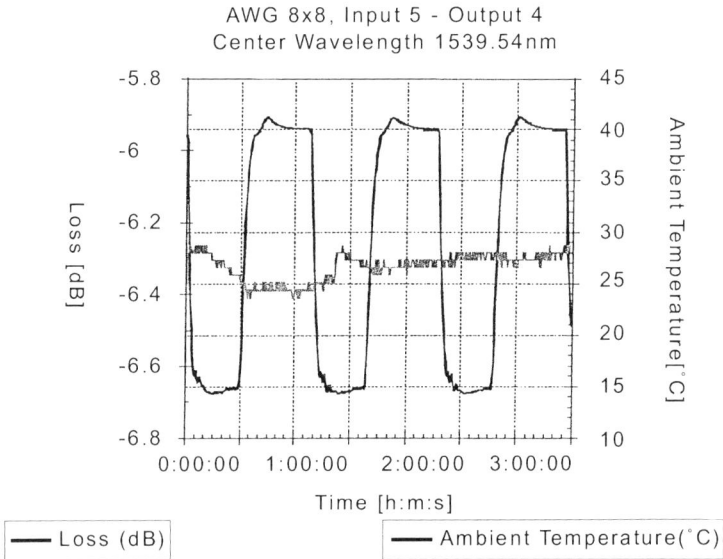

AWG 8x8, Input 5 - Output 4
Center Wavelength 1539.54nm

Figure 6.17 Temperature cycling of 8 × 8 module; $T_{chip} = 23$ °C. Reprinted with permission from Ehlers et al., *Optical Fiber Tech. 6*, 344 (2000). © 2000, Academic Press.

the joint, a glass plate was positioned onto the chip directly in face with the array and fixed there as shown in Figure 6.16. After curing, the opposite chip interface was aligned and connected in the same way as described previously using the already fixed fibers for launching the input wavelengths. To assess the alignment and stability of the system, several environmental tests were carried out, including a vibration test. In all cases, the insertion losses, center wavelength, and return losses were measured online at a constant chip temperature of 23 °C. The module was subjected to thermal cycles between 15 and 40 °C (not as severe as the Telecordia requirements), and the variation of the coupling loss was less than ±0.1 dB (see Figure 6.17). Finally, the vibrational tests were performed with an acceleration greater than 16*g* within a broad spectral range of 50–5000 Hz. After these tests, no significant degradation (<0.2 dB) of the coupling efficiency was detected. No high-temperature/high-humidity testing was performed on these devices in that work but would have to be done to qualify the device for telecommunications applications.

The next case relates to UV polymer replicated optics. A growing application area is found in the manufacturing of precision diffractive and refractive micro-optical elements such as lens arrays and polarizers. Replication of accurate masters in UV-cured polymers is one of the few cost-effective manufacturing methods that can meet the demand for low-cost yet high-quality micro-optical components produced in large volumes. In the replication by UV light, the microstructures are copied into a thin film of UV-curable epoxy (see Figure 6.18). The substrate can consist of a standard glass plate as well as high-precision-machined refractive optical elements such as lenses or prisms. The thickness and uniformity of the replicated film can be controlled to form an overall surface planarity of fractions of a micrometer over areas of some millimeters, thus being comparable to the quality of etched components in fused silica. The replication process is dependent on low shrinkage of the polymer on cure or at least a very controlled shrinkage on cure. It would be interesting to apply the principle of Step Cure to this UV-embossing process to control the shrinkage. The UV-embossing process has been extended to replicate micro-optical elements directly on optoelectronic components and detectors. Aligned double-sided replications that are possible with UV embossing are of great interest for data communications applications. UV-embossed elements can be replicated on a wafer scale and subsequent dicing can lead to a very competitive price for the individual elements.

At this juncture, it is important to address the issue of the need for a common test vehicle, especially for the assembly of components and systems for the fiber-optic telecommunications market. In the previous discussion, adhesive properties were considered by measuring sample coupons that certainly bear no relationship to the fiber-optic components. Indeed, several thermally cured materials have been qualified by fiber-optics companies, even though test coupons

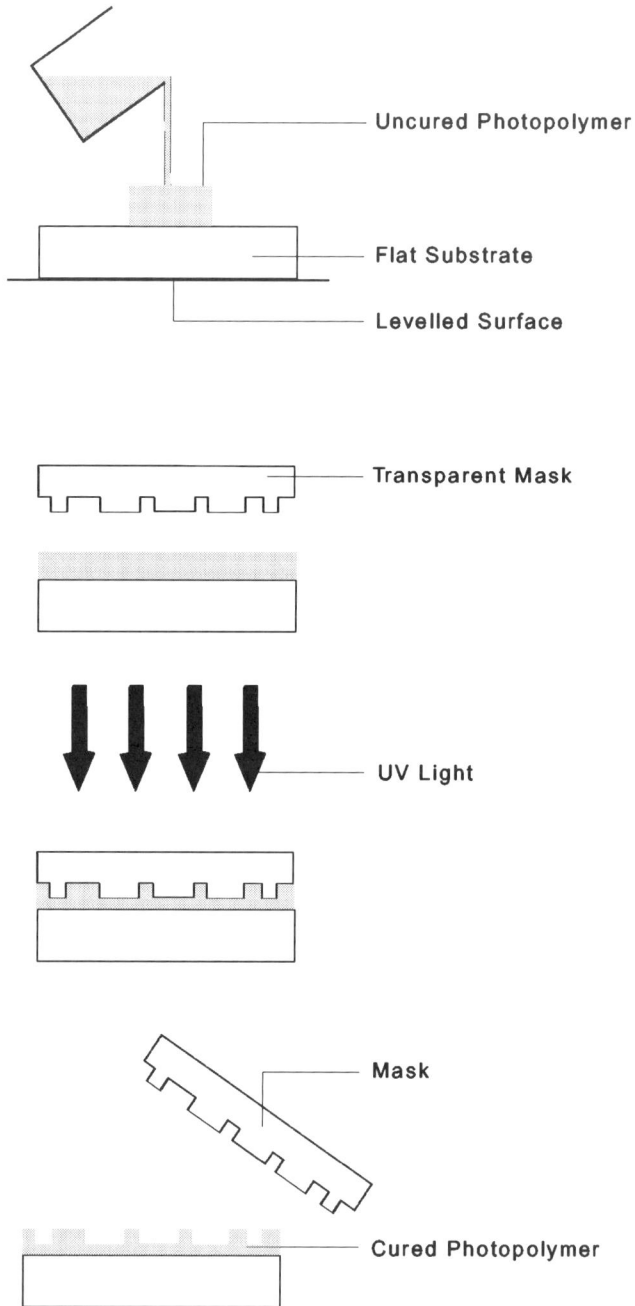

Uncured Photopolymer

Flat Substrate

Levelled Surface

Transparent Mask

UV Light

Mask

Cured Photopolymer

Figure 6.18   Schematic illustration of UV molding process.

made from the adhesive failed the environmental testing. This means that the adhesive behaves differently as a small blob than it does as a macro blob, or the design of the device reduces the impact of the adhesive sensitivity to environmental conditions due to joint design. How does one (or can one) determine that the optical components made with the adhesive bonds are stable? In this context, the most common approach is to (i) design the bond joints correctly, (ii) choose an adhesive based on previous experience or manufacturer's recommendation, (iii) build a prototype optical component, (iv) tweak the optomechanical design if the component changes with time, and (v) reiterate the design until movements are reduced to acceptable levels. These are labor-intensive and expensive methods that do not always produce appropriate results. Therefore, it would be preferable to have a test vehicle that more closely resembles the fiber-optic device and that is relatively cheap, easy to produce, and sensitive to environmental conditions that would allow for assessing the stability of individual bonds in a convenient and quick way. Stability is defined by the degree of change in the coupling efficiency as a function of alignment and environmental conditions. Daly (2002) describes a test jig, consisting of a mirror bonded to a mechanical mount, that can be put into an environmental chamber with a window for optical access. The mirror is aligned using an autocollimator and monitored as a function of change in the environmental conditions such as high dry heat, thermal cycling, and damp heat. The autocollimator has microradian resolution and is quite sensitive to angular changes. Although this provides some guideline to adhesive performance, it still does not approximate a device and is time

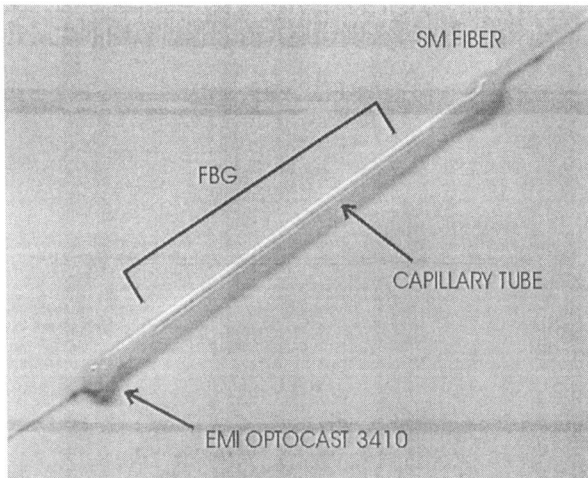

Figure 6.19 Fiber Bragg grating package test vehicle. Reprinted with permission from Martin and Hubert, *Proceedings of SPIE* (2002). © 2002, SPIE.

Fiber Bragg Grating - 1
Stress vs Wavelength Shift

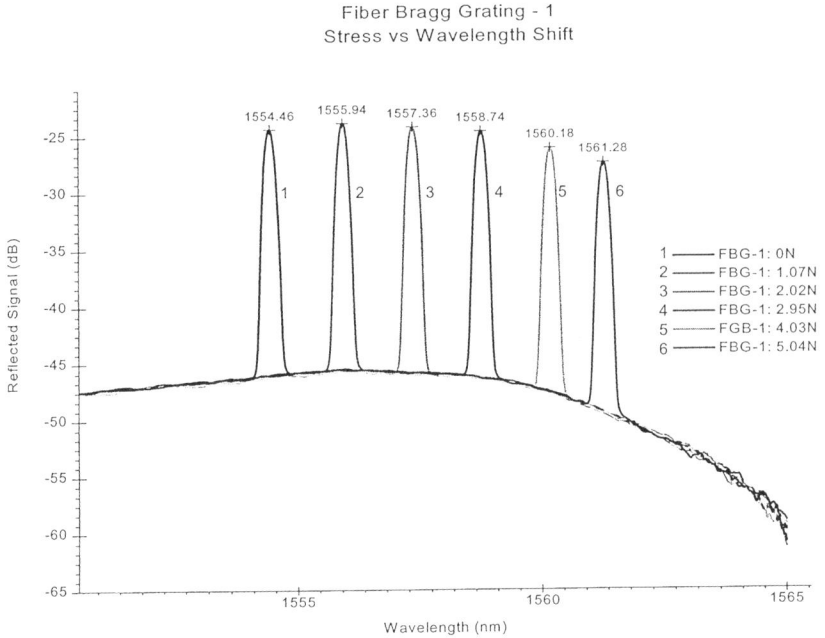

Figure 6.20 Shift in center wavelength versus stress. Reprinted with permission from Martin and Hubert, *Proceedings of SPIE* (2002). © 2002, SPIE.

consuming. Another approach has been taken by Martin and Hubert (2002) and Schulz et al. (2000). They have used devices incorporating fiber Bragg gratings to develop test vehicles. Martin and Hubert (2002) have packaged fiber Bragg gratings under tension into a capillary tube (see Figure 6.19) and cured the adhesives with different Step Cure profiles. The grating is held at each end of the capillary using an adhesive such as EMI 3410. The center wavelength, which varies as a function of the applied tension (see Figure 6.20) of the grating, is recorded with an optical spectrum analyzer. The device is then subjected to temperature cycling, damp heat, and so forth. The shift of the center wavelength is a measure of a decrease or increase in the tension on the Bragg grating due to shrinkage or degradation of the adhesive, as shown in Figure 6.21. Schultz et al. (2000) follow the same procedure, except they use a polarization-preserving fiber and they write two overlaid gratings (see Figure 6.22). This results in four peaks, two for each polarization and grating. If this fiber is subjected to an axial strain, the two peaks will shift together; if it is subjected to a transverse strain, the separation of the two peaks will change (see Figure 6.22). By embedding these fibers into a joint and with proper orientation, the stress of the joint can be monitored as a function of time or environmental conditions.

Fiber Bragg Grating

Shift after 100 Hours at 95C

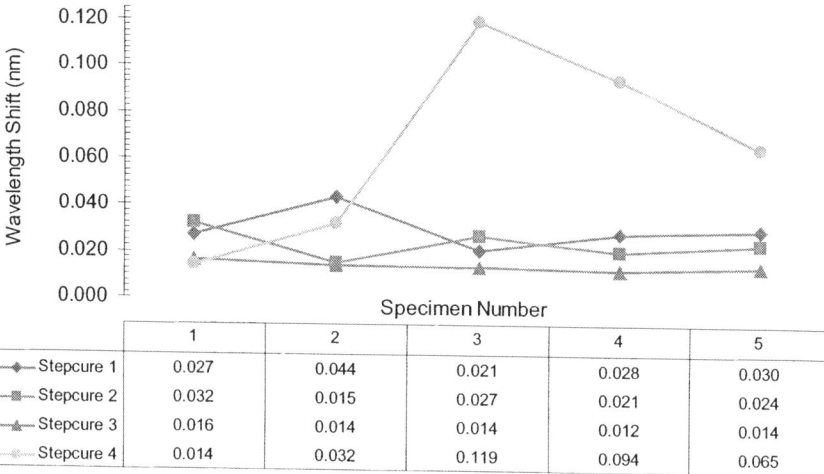

	1	2	3	4	5
Stepcure 1	0.027	0.044	0.021	0.028	0.030
Stepcure 2	0.032	0.015	0.027	0.021	0.024
Stepcure 3	0.016	0.014	0.014	0.012	0.014
Stepcure 4	0.014	0.032	0.119	0.094	0.065

Figure 6.21 Shift in Bragg grating center wavelength as a function of step-cure and temperature cycling. Reprinted with permission from Martin and Hubert, *Proceedings of SPIE* (2002). © 2002, SPIE.

Recently, Bourne and Thadani (2002) developed a device that can be used for testing adhesive bonds using white-light interferometry. Their apparatus is shown in Figures 6.23 through 6.25. They have published data for several UV-cured and thermally cured adhesives, including EMI 3411, Epoxy Technology 353, and Epoxy Technology OE188. These results are presented in Figure 6.26, which shows the cumulative movements after the bonds were cycled through a series of environmental tests, including thermal soak, thermal shock, thermal cycling, and vibration testing. A universal testing jig would be a significant advance in developing effective adhesive materials for alignment-sensitive devices such as single-mode fiber or wave-guide devices for long-term telecommunications applications and would substantially reduce the cost to manufacture such components.

6.4. ADHESIVE BONDING IN OPTOMECHANICAL SYSTEMS

As mentioned in Chapter 1, optomechanical adhesive bonding has more stringent requirements than other adhesive bonding applications. This is due to the fact that the prevention of an optical distortion requires minimization of stress,

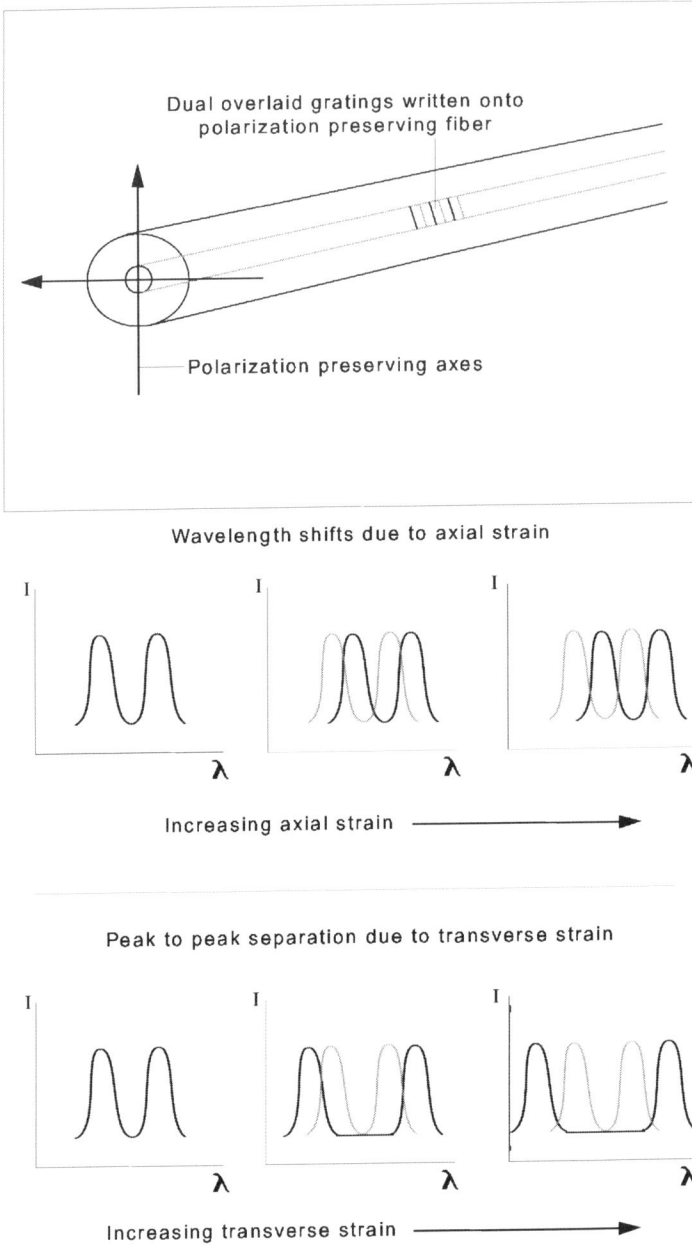

Figure 6.22 Multiaxis fiber grating strain sensor for measuring axial and transverse strains and response of multiaxis fiber grating strain sensor to axial and transverse strains. Reprinted with permission from Schulz et al., *Proceedings of SPIE*, 3991 (2000). © (2000) SPIE.

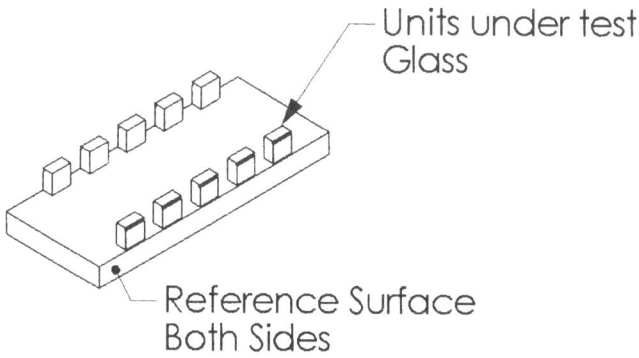

Figure 6.23 Test samples bonded to a BeO substrate. Reprinted with permission from Bourne and Thadani, *Proceedings of SPIE*, 4771 (2002). © 2002, SPIE.

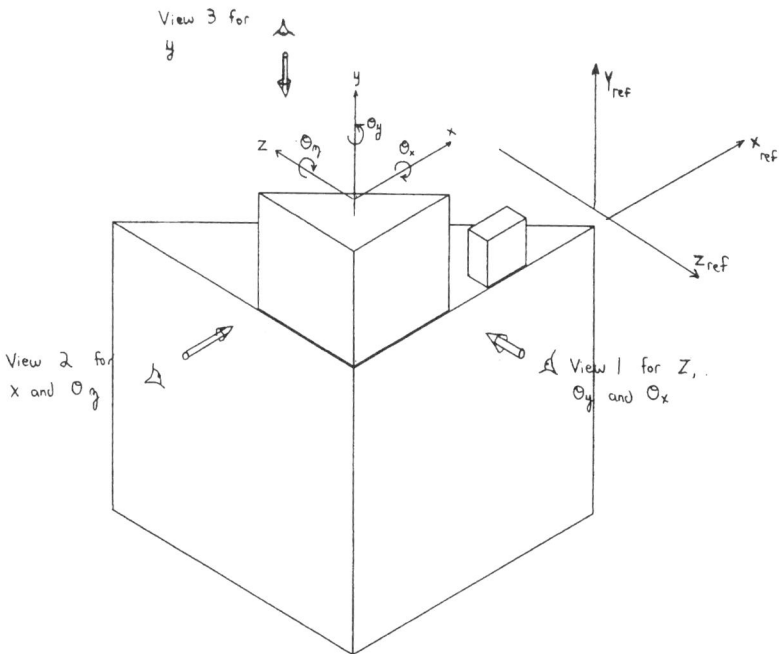

Figure 6.24 Test samples bonded to a reference substrate. The large prism is the reference object with three of its surfaces used as such. The test pieces are bonded to the reference object. Reprinted with permission from Bourne and Thadani, *Proceedings of SPIE*, 4771 (2002). © 2002, SPIE.

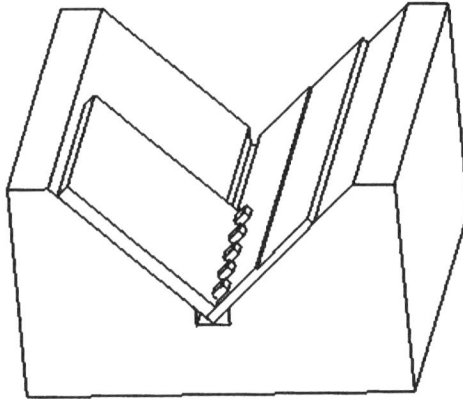

Figure 6.25   Test jig. Reprinted with permission from Bourne and Thadani, *Proceedings of SPIE,* 4771 (2002). © 2002, SPIE.

with adhesively bonded components remaining precisely in position throughout their operating lifetime. In this context, among the physical properties required for optimum optomechanical performance, the most crucial ones include adhesion, shrinkage, elongation, and modulus of elasticity. It is very difficult to optimize all these properties simultaneously for a given system, and, typically, a compromise between different properties is established.

In general, the lowest stress and greatest long-term stability requirements can be realized by minimizing the shrinkage. Although improved adhesion is required to withstand temperature cycling and differential expansion, a more resilient product having a lower modulus with higher elongation may be more advantageous to cope with these conditions. The maximum temperature range is essentially determined by the glass transition temperature T_g, which can be vital for rigid materials, as a rapid change in properties could influence their performance.

Typical optical components that can be mounted by employing UV adhesive mounting are lenses, prisms, mirrors, as well as fiber optics. In such cases, only relatively small (and low-mass) components up to about several centimeters in diameter can be used, which is related to obtaining adequate bond line thickness that can be cured with minimal shrinkage and stress.

Structural adhesives are also employed to attach and to seal various micro-optical and micromechanical components. One should emphasize again that joining such components requires more stringent requirements related to ensuring minimal stresses in the joints and the positioning accuracy of the components. In the case of the performance of devices such as MEMS (micro-electro-mechanical system), an additional issue is that of thermally induced

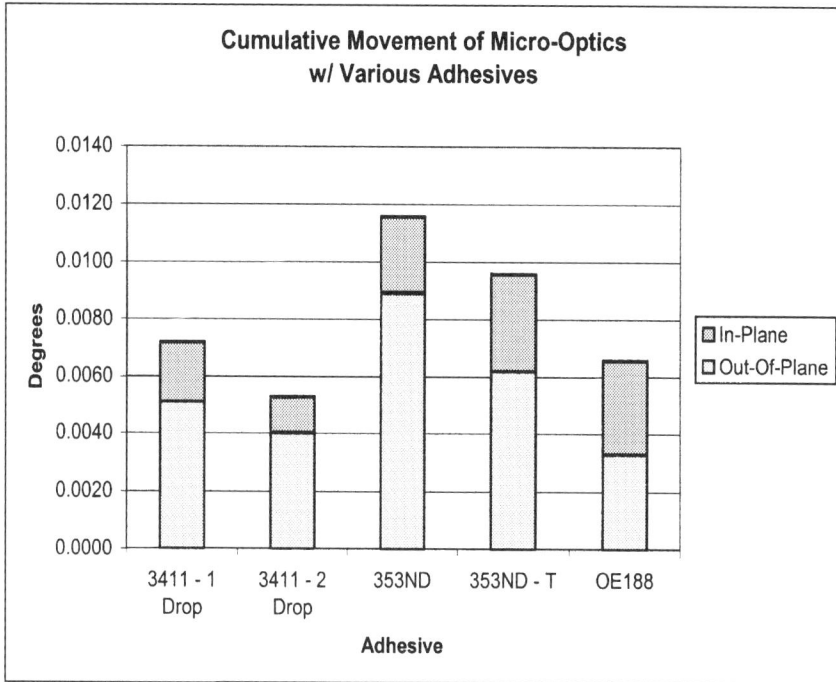

Figure 6.26 Cumulative movements in both in-plane and out-of-plane directions. Reprinted with permission from Bourne and Thadani, *Proceedings of SPIE*, 4771 (2002). © 2002, SPIE.

stresses. Considering also the need for long-term stability, MEMS-based devices present an important challenge to the assembly and packaging processes. In such cases, a package typically includes a range of technologies, such as motion, electronics, optics, chemistry, and biology. Some of the issues related to the packaging of such systems concern the facts that in specific cases the package must be hermetic, whereas some systems necessitate electrical and optical input and output, and in other applications minute amounts of fluids flow in and out of the structure.

Some of the critical cure considerations will be briefly outlined below. An important factor is the speed of cure, which depends mainly on the light intensity. It should be emphasized that the higher intensity of UV light can provide a faster cure, but it can also result in greater stress. Employing a lower intensity of light, on the other hand, leads to slower (gradual) cure and shrinkage, which may reduce stress. Another critical issue is that related to the uniformity of illumination that avoids shadows over the bond line. The choice of the most suitable adhesive in these applications involves an evaluation of such important properties as adhesion to substrates and shrinkage, as well as modulus and elongation. In this case, it is important to ensure that the adhesive has the appropriate trade-off in materials characteristics (see Section 2.5).

6.5. SUMMARY

Having various sizes and tolerances, optical, fiber-optic, and optoelectronic components and devices are assembled into a package designed to include the coupling of the electromagnetic radiation into and/or out of optical fiber cables. Such components are commonly joined with adhesives that provide great flexibility and versatility for joining dissimilar materials. To avoid any power loss at the interface, optical adhesives are selected so that their indices of refraction are matched to the indices of refraction of the optical fibers. Typically, optical radiation curing of adhesives is a preferred method, providing high cure speeds and throughputs.

In photonics applications, especially the fiber-optic telecommunications industry, the precision prealignment considerations are very stringent, when compared to the use of adhesives for microelectronic applications. Such issues as maintaining alignment and its long-term stability under severe environmental conditions, directed at minimizing any transmission losses of light at the joint, are of vital importance.

The main advantages of using photocured adhesive bonding, which include speed of attachment, uniform distribution of loads, and the ability to bond dissimilar materials, can be compromised if the adhesive is not effective in providing a "seal" against the environment. Adhesive materials can exhibit shortfalls in physical properties for precision-aligned structures, most notably by misaligning the assembly through the shrinkage that accompanies polymerization or by weakening of the bond under high-temperature and humid conditions. These effects can be ameliorated by judicious bond design and by controlling the curing of the adhesive with proper curing programs and beam delivery design. A major advance to removing these problems with adhesives would be the development of a good test vehicle, which would allow the monitoring of the stability of the adhesive bonds under the appropriate environmental conditions.

CHAPTER 7

Issues Related to Optical Adhesive Bonding

CONTENTS

7.1. INTRODUCTION

In general, there are two basic viewpoints in evaluating an adhesive joint. One view, which is related to the interface properties of the adhesive bond, attributes any changes in the adhesive joint to modification of the chemical bonding at the interface. The other view is based on the mechanical behavior determined by the bulk properties and geometric characteristics of the adhesive joint.

The identification of techniques and methodologies for evaluating the failure mechanisms of adhesive bonds is of great importance. In this context, the main emphasis is directed toward developing a method (or methods) for analyzing and predicting failures in adhesive joints.

It should be emphasized that the understanding and control of interface formation between dissimilar materials is a complex issue. There are many factors that may affect adhesion during the manufacturing process (see, e.g., Pignataro, 1998). These may include, for example,

(a) Chemical bond (determined by interatomic and intermolecular forces)
(b) Interfacial contamination
(c) Induced modification of an interface
(d) Mechanical interlocking (determined by interpenetration at both nanometer and macroscopic scales)
(e) Stability under working conditions (effect of high temperatures and high humidity)

The processes related to the interfaces (e.g., contamination, diffusion, reaction between various species, and induced modification) determine the properties of practical interfaces and the tendency of the adhesive bonding toward degradation. Finally, the interface properties, and thus adhesion, may undergo modifications with time under working conditions (see, e.g., Pignataro, 1998) due to temperature and humidity excursions.

7.2. FAILURE MECHANISMS

There are two principal mechanisms of failure associated with adhesive bonding: adhesive failure and cohesive failure. Adhesive failure is the failure at the interface between the adhesive and one of the adherends (this is typically due to a weak boundary layer arising from, for example, inadequate surface preparation or unsuitable adhesive choice). A major difficulty associated with analyzing adhesive failure in fiber-optic components is the variability of the manufacturing process, including assembly, cleaning, curing intensity, and so on. Cohesive failure is associated with the internal breakdown of either the adhesive or, rarely, one of the adherends. These failure mechanisms are shown in more detail in Figure 7.1. Typically, the adhesive bond failure is neither completely cohesive nor fully adhesive. The exact origin of the premature failure in adhesively bonded joints is very complex to determine. This may include, for example, undesirable stresses, temperature, humidity, fatigue, solvents, and variable processing conditions.

The performance requirements for adhesives used for bonding between optical and mechanical components are more stringent when compared to other

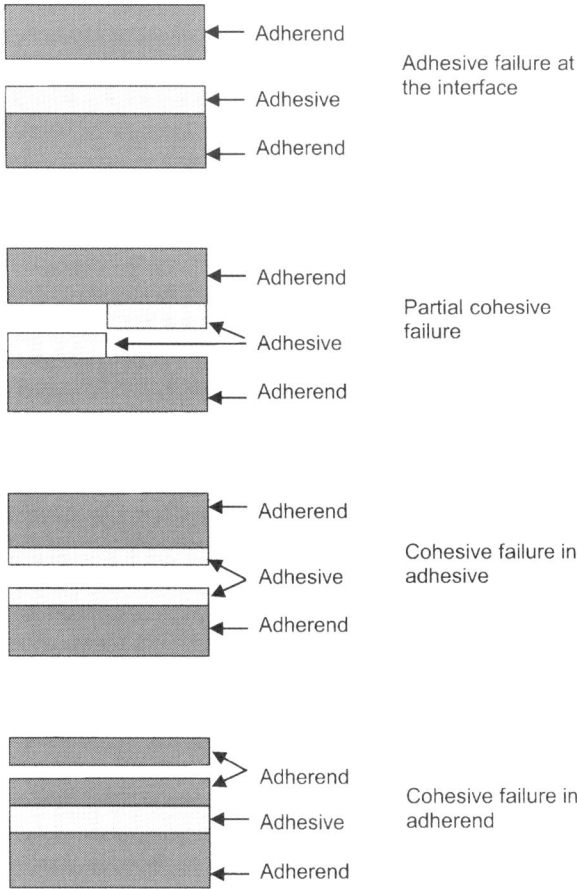

Figure 7.1 Schematic (idealized) diagrams of different types of adhesive and cohesive failures.

bonding applications. In order to avoid optical distortions, stresses must be minimized over the complete range of operating temperatures and bonded components must maintain tight alignment tolerances throughout the operational lifetime of the systems.

In general, various interrelated causes may influence the reliability of the adhesive bond (Woods, 1993). The cause-and-effect diagram (also referred to as the Ishikawa diagram) can be used, for example, for assessing various causes of failure in optical assemblies employing UV curing adhesives and for providing a framework for the analysis of the important relevant process variables (Woods, 1993). In this context, the main issues relate to (i) the selection, properties, and handling of the adhesive in relation to the design requirements (physical properties of importance are coefficient of thermal expansion, viscosity, refractive index, modulus of elasticity, tensile strength, hardness, elongation at failure, temperature range); (ii) the physical properties, design characteristics, and fabrication method of the optical components; (iii) UV sources; (iv) cleaning; (v) cure; and (vi) assembly environment.

7.3. STRESSES AND DEGRADATION

Cure stress is defined as a residual internal stress originating between different components (due to different coefficients of thermal expansion) during the curing process, whereas degradation is defined as a detrimental modification in physical properties, chemical structure, and/or appearance.

In many cases, manufacturing problems are related to substantial stresses generated during the curing process of adhesives. Thus, it is of great importance to be able to predict the development of such stresses, which can be related to changes in viscoelasticity during the curing process.

There is a need for the evaluation of thermal and mechanical stresses in microelectronic and fiber-optic assemblies. Such stresses include (i) thermally induced stresses in an optical fiber in a ferrule, (ii) the mechanical interfacial shearing stress in a fiber mounted in a ferrule and subjected to tension, and (iii) stresses in a fiber-optic interconnect experiencing bending.

Structural analysis, as well as the analysis of thermally induced stresses in microelectronic and fiber-optic systems, was carried out extensively by Suhir, who derived analytical expressions for the evaluation (and prediction) of the thermal stresses in various adhesively bonded assemblies (Suhir, 1991, 1999, 2000, 2001). Such analyses are very useful in the stress–strain analysis and physical design of various microelectronics and photonics structures and devices.

One of the crucial issues in achieving optimum optical performance of fiber-optic connectors is good physical contact at the interface between two fibers. Some of the issues of concern are related to the material degradation of (i) the fiber, (ii) the ferrule, and/or (iii) the adhesive that actually holds the fiber into the ferrule. The ability to monitor and to quantify mechanical variations in the materials, such as adhesive creep, ferrule deformation and roughening, and fiber deterioration, is of great importance.

One of the important issues is also related to minimizing the detrimental effect of voids, present in the adhesive, on the thermally induced stresses in optical fibers embedded in the adhesive. In such cases, the problem may arise due to the stress concentration resulting in the origination of cracks in the epoxy.

An important issue is how to minimize stresses generated during the curing process. This would require a basic understanding of the intricate processes related to changes in viscoelasticity as the cure proceeds. Such understanding would allow, in principle, determining a cure timetable that would result in minimized stress generation during cure. The modeling of the evolution of stresses in cross-linking polymers facilitates the development of practical ways to minimize stress through control of the cure cycle (Adolf and Martin, 1996). It should be noted, however, that stress modeling is rather challenging due to complex calculations and the fact that such material characteristics as, for example, stress versus strain curves, Young's modulus, and the Poisson ratio for relevant materials are not readily available (Swanson and Enlow, 1999).

7.4. ALIGNMENT DISTORTION DUE TO CURING

The distortions of the previously aligned components (of photonics assembly) induced by curing and cooling are of great concern in photonics assembly. This issue has been investigated by Lin et al. (2002) and Zhou et al. (2002), who determined the extent of the distortion (of the previously aligned components) produced by the curing and cooling processes of the adhesive, and correlated the misalignment with the adhesive properties and bonding structure. Such distortions typically result in reduced assembly coupling efficiency. This problem could be alleviated by appropriate design considerations for an adhesive joint. Lin et al. (2002) have analyzed adhesive-curing-induced misalignment in fiber-optic assembly of an eight-channel collimator array by employing a three-dimentional (3D) finite-element analysis method for investigating the effects of (i) the materials properties of an adhesive, (ii) adhesive joint parameters, and (iii) UV curing conditions on the relative displacement (i.e., misalignment of fiber and lens arrays) during the curing process. It was concluded that, for a given adhesive and curing conditions, the adhesive joint could be designed to alleviate the alignment distortion during curing and cooling (Lin et al., 2002). Specifically, it was deduced that choosing UV-curable adhesives with low shrinkage ratio and low Young's modulus and using a suitable adhesive joint design result in (i) reduced adhesive-curing-induced relative displacement in the optical assembly and (ii) improved coupling efficiency. It was also concluded that the rough adhesive surface, which may affect the coupling loss, should be avoided during the UV curing process. For the evaluation of optical misalignment tolerance, Lin et al. (2002) investigated a 1×8 collimator array, consisting of a 1×8 fiber array and a 1×8 lens array, having the following dimensions (in millimeters): fiber array, $10 \times 10 \times 1$; lens array, $1 \times 10 \times 2$; and glass plate, $5 \times 10 \times 0.2$. During the assembly process of the collimator array, the curing-induced adhesive shrinkage results in substantial relative movements between the fiber and lens arrays, and hence this leads to misalignments between the fiber and lens arrays and loss of coupling power.

The basic conclusions of the investigations by Lin et al. (2002) are as follows:

(i) Using UV-curable adhesives with lower shrinkage ratio results in less curing-process-induced displacement in the optical assembly, lower stresses, and improved coupling efficiency in photonics assembly employing adhesive bonding.

(ii) The adhesive layer thickness has a significant effect on the deformation of the optical assembly. Appropriate design of the adhesive joint can alleviate the displacement induced by UV curing and cooling processes on the optical loss. Specifically, it was determined that, regardless of curing shrinkage, a proper choice of bond line thickness can reduce the curing-induced coupling loss.

(iii) An uneven surface in the adhesive layer and/or nonuniformity of the adhesive material influence the relative displacement of the fiber array and lens array. Thus, these factors have to be avoided during the curing process.

(iv) The Young's modulus of adhesive has a significant effect on both the displacement and the stress distribution in the adhesive layer. The relative displacement and the stress in the adhesive layer increase with increasing Young's modulus, resulting in greater misalignment of the assembly. Thus, the selection of an appropriate adhesive with the lowest possible Young's modulus is of great importance.

7.5. CHARACTERIZATION METHODS

7.5.1. INTRODUCTION

A wide variety of characterization techniques are available for failure analysis and quality control of adhesive bonding. Failure modes are often associated with manufacturing process-induced flaws or with imperfection-dependent degradation in service, which may also be interdependent. The analytical techniques can typically provide complementary information related to the material's physical, structural, and device properties (see, e.g., Brundle et al., 1992).

As mentioned above (in the section on real-time monitoring of adhesive curing), some characterization techniques can be employed *in situ* (i.e., during the material's preparation or processing), whereas others can only be used *ex situ* (i.e., analysis is performed after the preparation or processing of the material or device). It should be emphasized that, in the case of adhesive curing, it is of great advantage to use *in situ* techniques, since they offer the analysis of a process in real time. (It is important to note that the adhesive must be prepared properly, i.e., mixed adequately, used at proper temperature, stored properly within expiration date, etc.)

7.5.2. THERMAL AND DYNAMIC MECHANICAL ANALYSIS METHODS

In addition to the characterization techniques outlined below, there are also some important methods for the analysis of adhesives. These include, for example, thermal analysis techniques such as (i) thermal gravimetric analysis (TGA), which monitors the sample weight changes as a function of temperature; (ii) differential thermal analysis (DTA), which monitors changes in the enthalpy of the materials as a function of temperature; (iii) differential scanning calorimetry (DSC); and (iv) thermal mechanical analysis (TMA) in conjunction with DTA to derive the thermal transition properties such as the glass transition temperature T_g. Other important analytical techniques are gas chromatography (GC) and GC mass spectroscopy, as well as combined thermal analysis and GC mass spectroscopy for analyzing volatile reaction products during cure. These techniques have previously been outlined in Section 5.5.2.

7.5.3. MICROSCOPY

Various microscopy techniques (providing different spatial resolutions) are available for the analysis of a wide range of materials properties.

The electron microscopy techniques that can provide information related to adhesive bonding include scanning electron microscopy (SEM) and transmission electron microscopy (TEM).

The *scanning electron microscope* (SEM) is a multimode instrument. These modes, providing complementary information on the structure, composition, and physical properties of materials and devices, make the SEM one of the most indispensable tools in materials characterization. The SEM can accommodate and examine macroscopic specimens (e.g., whole structures or assemblies) with little or no special specimen preparation. The use of the secondary electrons in SEM provides the most routinely used *secondary electron image* (SEI) mode for observing the topographic features of solid surfaces. Under appropriate operating conditions, *backscattered electrons* provide crystallographic information due to electron channeling, which is based on the fact that the backscattered electron yield varies as the angle of incidence of the scanned electron beam passes through the Bragg angle to the crystal lattice planes. *Characteristic X-rays*, emitted due to electronic transitions between inner-core levels, can be used to identify the particular chemical element present and its concentration, that is, the composition.

In *transmission electron microscopy* (TEM), electrons are transmitted through thin samples (on the order of 100 nm and less) and focused to an image by further electron optical lenses in a way completely analogous to the (transmission) light microscope. This provides information related to the crystal structure and defects in solid-state materials. The crystalline structure can be determined by using *transmission electron diffraction* patterns, which can be formed by inserting the *selected-area diffraction* (SAD) aperture in the image plane of the objective lens to give selected-area diffraction patterns. One of the powerful TEM techniques is the cross-sectional TEM (XTEM), which provides a powerful means for the analysis of the interfaces in multilayer structures; however, the sample preparation in this case is fairly tedious, destructive, and time consuming.

Scanning acoustic microscopy (SAM) is based on the generation and detection of elastic waves in solid materials. In this technique, the interaction of an acoustic wave with the material is basically dependent on the mechanical properties of the material. Acoustic waves can penetrate materials that are opaque to optical or electron-beam irradiation; thus, SAM is capable of imaging subsurface features and local variations related to mechanical properties, such as density, elasticity, and viscosity. Specifically, this technique is suitable for the examination of defects such as cracks, voids, and inclusions.

The development of *scanning probe microscopy* (SPM) techniques was catalyzed in the last decade by the invention of the *scanning tunneling microscope* (STM), which is capable of imaging surfaces on the atomic scale (see, e.g., Wiesendanger, 1994). The operation of the STM is based on the measurement of the electron tunneling current between an ultrasharp tip and the sample. The tip is made of conductive material, and it can be moved in three dimensions by employing piezoelectric elements for x, y, and z translators. The tip is positioned about 10 Å above the sample surface, so that, at an operating potential difference on the order of millivolts, a tunneling current of about 1 nA is detected.

Another SPM technique, *atomic force microscopy* (AFM), can be employed in the analysis of both conductors and insulators. In this case, a tip, such as a diamond crystal fragment, is attached to a flexible cantilever that is deflected due to the interaction force between the tip and the sample surface. The atomic interaction force, experienced by the tip, can be derived from the deflection of the cantilever that can be measured employing either electron tunneling or optical detection. The deflection, and thus the interaction force, is controlled by a feedback system, which allows recording of the topography of the sample surface. A common interatomic force that contributes to the deflection of the cantilever is that due to the van der Waals interaction. This is repulsive in *contact mode* (i.e., with a tip-to-sample separation less than a few angstroms) and attractive in *noncontact mode* (for a tip-to-sample separation in the range between about 10 and 100 Å). Atomic force microscopy is especially useful in studies of the fundamental mechanisms of adhesion and in relating the adhesive properties of materials to their nanoscale surface structure (e.g., Burnham and Colton, 1989; Colton et al., 1998; Somorjai, 1998; Serry et al., 1999). Using the so-called lateral force microscopy (or a frictional force microscopy) one can also elucidate the two-dimensional nature of the atomic-scale friction, which can be explained in terms of the so-called stick–slip model (Fujisawa et al., 1995).

Another promising technique for investigating the surface properties of materials is *chemical force microscopy*, incorporating the tip (which is functionalized with specific chemical species facilitating the interpretation of measured forces in terms of specific interactions between the molecules attached to the tip and the molecules present on the surface), which is scanned over a sample in order to determine adhesion and friction forces between the given chemical species on the tip and those on the sample surface (see, e.g., McKendry et al., 1999).

For analysis of the optical properties of materials, the basic principles of SPM have also led to the development of the technique known as *scanning near-field optical microscopy* (SNOM), which allows one to achieve a spatial resolution exceeding the far-field diffraction limit by removing lenses completely from the imaging system and employing the near-field collimation of light. The resolution is limited by the size of any aperture in the system and can be as small as several nanometers.

The mechanical properties of thin films can also be derived from nanoindentation measurements (see, e.g., Adhihetty et al., 1998). It is also, in principle, feasible to evaluate the thin-film interfacial adhesion by employing the cross-sectional nanoindentation technique (Sanchez et al., 1999); however, this was applied to ceramic materials (when applied to adhesives, plasticity has to be considered, which would make it difficult to interpret). It would be desirable to employ this technique to measure the profiles of the mechanical properties across the interface and even to map out the distributions of various mechanical properties. A viscoelastic analysis of nanoindentation has been applied for the interpretation of the nanoindentation tests in relation to the mechanical properties of polymeric coatings (Cheng et al., 1998, 2000; Grau et al., 1998; Strojny and Gerberich, 1998). Such studies provide a basis for investigating the elastic and viscous properties of coatings based on microindentation and nanoindentation experiments.

Note, however, that the volumes or areas analyzed, using the above microscopy techniques, are typically relatively small. Therefore, the results may be highly atypical unless the observations are repeated in different regions.

7.5.4. SPECTROSCOPY

Raman spectroscopy is one of the most promising methods for the analysis of curing processes related to adhesives. This method is based on measuring the energy shift of the incident photon beam that is inelastically scattered off the material. In this measurement technique, the specimen is illuminated with a monochromatic light generated by a laser, and incident photons, inducing transitions in the material, gain or lose energy. The energy shift during such a scattering process is due to either the photon energy transfer to the lattice (i.e., phonon emission) or the absorption of a phonon by the photon. In the case of phonon emission, the reduction in photon energy is called the Stokes shift, and in the case of scattered photon emerging at a higher energy, it is called the anti-Stokes shift. Raman spectra contain information on the vibrational modes in the material. The coupling of an optical microscope with the Raman system also allows one to obtain spatially resolved Raman measurements, that is, *micro Raman spectroscopy*, with a resolution of about 1 μm. In this case, the illumination of the sample by a laser is through an optical microscope coupled with the monochromator. Raman spectroscopy allows distinguishing between crystalline and amorphous materials from the variations in the shift, the width, and the symmetry of the Stokes line and can also provide important information on defects, induced damage, stresses, and materials processing. Some of the main advantages include the fact that this is a nondestructive technique that requires no vacuum. In fact, Raman spectroscopy is one of the most promising techniques for real-time monitoring of the curing process.

Infrared absorption spectroscopy can be used to determine molecular bonding and coordination. Molecular bonds, described as spring oscillators, have resonant frequencies related to stretching or bending vibrational mode. Typical values of the resonant frequencies are on the order of 10^{14} s^{-1}, which corresponds to a wavelength λ of about 3 μm, that is, in the near infrared. (In infrared spectroscopy, a more convenient quantity used in the analysis is a wavenumber, which is the reciprocal of the wavelength, expressed in units of cm^{-1}.) The characteristic frequencies of various bonds can be identified from the tabulated data given in the infrared spectroscopy literature (databases). Thus, infrared absorption spectroscopy can be used, for example, for the identification of various resin components. Typically, the spectral ranges of infrared spectrometers are between about 200 and 4000 cm^{-1} (corresponding to wavelengths between about 50 and 2.5 μm).

7.5.5. SURFACE AND INTERFACE ANALYSIS

A detailed knowledge of the surfaces and interfaces is crucial in adhesives technology. The most commonly used surface characterization techniques for adhesives are *X-ray photoelectron spectroscopy* (XPS), *Auger electron spectroscopy* (AES), and *secondary ion mass spectrometry* (SIMS). In most cases, these surface analytical techniques probe the top several monolayers of the material (with subsequent *in situ* etching of the surface with an ion beam, one can also derive bulk information and depth profiling of the property of interest), and they provide information about the composition, concentration, and distribution of various species in the material, as well as chemical information. In addition to these techniques, SPM techniques and SEM are widely used for the analysis of surface properties.

X-ray photoelectron spectroscopy (XPS) is one of the most important techniques for the analysis of adhesives. This technique, which employs X-rays for the excitation of the solid and the detection of emitted photoelectrons with characteristic energies, can provide chemical information about the material. The binding energies of the emitted photoelectrons are related to the photon energy $h\nu$ (in this case, X-ray), the kinetic energy, E_k, of the photoelectrons, and the work function of the sample, ϕ, following the relationship $E_B = h\nu - E_k - \phi$. Thus, from the measurement of the photoelectron kinetic energy, one can determine the electron binding energy, which is characteristic of the particular atom, and thus the corresponding atom can be identified. Due to quantized energy levels in the atoms, the photoelectron kinetic energy distribution is composed of a series of discrete bands. Since the energies of photoelectrons are much less than 1 keV, the escape depth, and thus the depth resolution, is within about 20 Å of the surface. Thus, in this technique, the energy spectrum of the photoelectrons, which are emitted from the sample, provides nondestructive elemental and chemical analysis of the surface. The main applications of photoelectron spectroscopy are in determining binding energies and in detecting particular elements present at the surface of the material. Since the atomic environment influences the binding energy of an electron, it is possible to obtain information on chemical bonding of a particular element from the *chemical shift*, which enables identification of the compounds. The main advantages of XPS are (i) its sensitivity, (ii) the nondestructive nature of the analysis, (iii) minimal sample charging and beam damage, and (iv) the capability to analyze chemical shifts from the same element in different compounds. The XPS technique is especially useful in elucidating interfacial interactions, which is one of the crucial issues in adhesive bonding; thus, the XPS technique can be effectively employed in the analysis of an adhesive failure involving, for example, increased concentrations of various species migrating to the interface region. For example, XPS can be used in elucidating the effects of laser-radiation-induced pretreatment on adhesion (Wesner et al., 1997) or the effects of plasma pretreatment of the substrate on improved adhesion (Bertrand et al., 1998); these studies relate adhesion to the chemical bonding at the interface as ascertained by XPS measurements. Such XPS measurements also provide important information related to the interface failure mechanisms (Jiang et al., 1998). It is also important to note at this juncture the usefulness of developing correlations between various interdependent properties, for example, the chemical changes occurring at the interfaces, as studied by XPS, and the mechanical strength of adhesion, as determined by delamination test methods (Senturia, 1995). Various measurements, including XPS, can also be combined to provide the correlation among chemical composition, crystallography, and adhesion at the interface (Wagner et al., 1997).

In *Auger electron spectroscopy* (AES), *Auger electrons*, which are emitted from about the top 10 Å of the material, have energies characteristic of the elements of the material. Because of the very short escape distance of the electrons, the techniques based on the detection of such electrons are inherently surface sensitive. Thus, *Auger electron spectroscopy*, providing analysis of surface compositional variations, constitutes an important surface analytical technique in ultrahigh vacuum electron probe instruments. AES is based on the detection of the electron that has been ejected due to the rearrangement of core electrons in the atom as the result of primary electron-beam bombardment with typical energies of about 5 keV. Using this technique, the binding energies of the core electrons in the atom can be deduced and the chemical elements can be identified. Auger electron spectroscopy can be combined with the slow removal of the outermost atomic layers of the material by sputtering employing argon ions; thus, AES can provide depth profiling of chemical constituents. By employing scanning Auger electron spectroscopy (i.e., the excitation electron beam is focused and scanned across a sample surface), it is also possible to obtain compositional images with high spatial resolution.

In *secondary ion mass spectrometry* (SIMS), the primary (incident) ion bombardment (e.g., O_2^+ or Ar^+ with energies of up to 20 keV) sputters off secondary ions of the material that are identified in a mass spectrometer. Two variations are *static SIMS* and *dynamic SIMS*. In static SIMS, the total primary ion dose is sufficiently low for the analysis of the surface monolayer only (however, signal levels are correspondingly low). In dynamic SIMS, the primary ion dose is sufficiently high to maximize the signal for trace element analysis but the sample surface is rapidly eroded, yielding depth-resolved information. Quantitative information is obtained by converting the ion count rates to atomic concentrations by using the *relative sensitivity factor*, derived from measurements on standards of known composition. One can also obtain SIMS images, and by obtaining a series of secondary ion images as a function of depth, it is also feasible to derive three-dimensional compositional information.

It should be noted that the combination of some of these techniques used in a single vacuum system is of great utility since they can often provide the most effective ways of obtaining relevant complementary information required to elucidate the problems related to adhesive bonding. Such combination of techniques include, for example, XPS/AES and XPS/SIMS.

7.6. MECHANICAL PROPERTIES AND TESTS RELATED TO ADHESIVES

Some of the critical issues related to adhesive bonding (and, in general, to various joint technologies employed in microelectronics and photonics) include the mechanical properties of bonded structures, such as fracture. Modeling of such properties (with abilities to estimate the strength of the adhesively bonded joints and to predict the deformation and fracture of the joints) and understanding the factors related to the fracture mechanics of adhesive joints are of great importance. There is extensive literature on this subject; for some recent articles, see, for example, Imanaka et al. (2000) and Yang and Thouless (2001) and references therein.

In the development of a new adhesive, some of the most important tests relate to (i) its strength, (ii) its fracture toughness, and (iii) the resistance of the bonded joint to various solvents, as well as to various environmental conditions of high temperature and humidity.

Typically, the strength of an adhesive bond is derived from destructive tests such as, for example, lap shear test, peel test, cleavage test, and double-cantilever beam test, which are usually conducted under various temperatures and environmental conditions. An adhesive bond can also be characterized by finding the energy required to cleave a part of the interface.

The typical test for strength characterization is the lap shear test; for fracture toughness, the double-cantilever beam test; and for assessment of the resistance to solvents, the wedge test.

Some of the geometries commonly used for testing adhesive joints are illustrated in Figure 7.2. For details on mechanical tests related to adhesive bond performance, see, for example, Adams and Wake (1984) and Pocius (1997). Additional tests required for some applications include, for example, those related to fatigue (i.e., cyclic loading) and creep (i.e., application of low stresses for prolonged times) (see Adams and Wake, 1984).

7.7. RELIABILITY ISSUES

Some of the most important issues in practical applications of adhesive bonding are those related to reliability, reproducibility, and prevention of any degradation throughout the operational lifetime of the assembled structure. Typically, the JEDEC (Joint Electron Device Engineering Council) JESD 72 and Telcordia G-1209-CORE, and G-1221-CORE testing requirements provide test methods and acceptance procedures for adhesives under the appropriate environmental conditions. These methods are usually laborious, expensive, and time consuming. Accelerated tests are being developed but must be carefully applied.

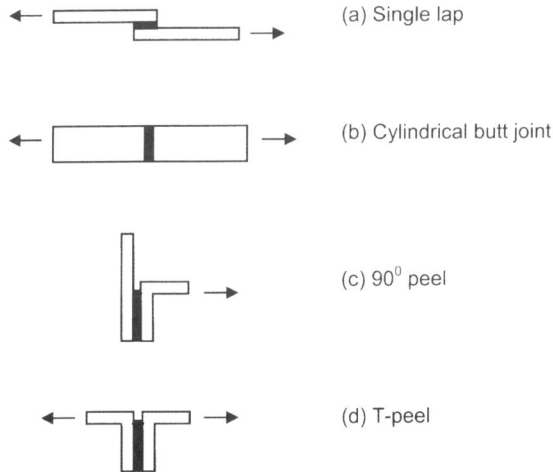

(a) Single lap

(b) Cylindrical butt joint

(c) 90⁰ peel

(d) T-peel

Figure 7.2   Some of the geometries commonly used for testing adhesive joints.

Furthermore, because of the fluid nature of polymers, they continue to flow or creep over long periods, especially if they are not optimally cured. This behavior can be monitored with appropriate test samples (structures or devices) that will be required to determine the long-term viability of using adhesive bonds for long-haul telecommunications applications.

The main question that arises in elucidating the failure of adhesive bonding is the issue of the sequence of analysis directed at extracting the causes of failure. In general, the issue of concern is whether it is possible to determine a dominant cause of failure without any prior knowledge of "input variables," that is, (i) surface contamination, (ii) surface roughness, (iii) preparation procedure for the adhesive mixture (it should be emphasized that after the adhesive is mixed and prepared it is typically very difficult to determine if there were any irregularities or flaws in its preparation), and (iv) curing conditions and their reproducibility (e.g., Was complete cure achieved? Was the process sufficiently well controlled? and, in this context, How does one define "complete cure"?). Thus, in the absence of real-time and continuous monitoring of all these variables, it would be difficult to fully ascertain the possible causes of failure.

One can apply a battery of tests using various analytical techniques, and one can determine a wide variety of existing properties of the adhesive and of the adherend, as well as determine the presence of possible interface contaminants. But the question will still remain as to the precise origin of the contamination. Or, one can determine the physical and chemical properties of the adhesive, but without having the whole history of the preparation or processing, can one determine the cause for the undesirable effects or characteristics? One can also determine (in a relatively straightforward way) the type of failure of the adhesive bond, that is, whether it is an adhesive or cohesive failure in the adhesive or in the adherend, or whether it is an adhesive failure at the interface, but it is very difficult to relate these types of failures to the "input parameters," that is, the characteristics of the material prior to adhesive formation or the detailed procedures of adhesive preparation, surface treatments, or curing conditions, unless all these are carefully monitored and characterized. Another example is that of stress failure, which may originate, for example, due to the residual stress produced during the curing process (e.g., the cooling process), or it could be related to the adhesive joint design.

Thus, to summarize briefly, unless all the "input variables" (or at least the prevailing ones) are carefully identified and monitored, it would be difficult to identify the precise cause for joint failure or for its degradation.

One of the important issues related to the reliability of UV-curable adhesives used in optical devices is the adhesive's resistance to degradation and dimensional and mechanical instabilities such as creep. In this context, it should be noted that although thermal postcuring may result in shrinkage and degradation, it might also raise the glass transition temperature of the adhesive, and thus improve the reliability related to mechanical instability.

7.8. SUMMARY

The two basic perspectives on evaluating an adhesive joint are related to (i) the interface properties of the adhesive bond and (ii) the mechanical behavior determined by the bulk properties and geometric characteristics of the adhesive joint. An important issue is the identification of techniques for elucidating the failure mechanisms of adhesive bonds, with the main effort being directed toward developing methods for predicting failures in adhesive joints.

The main factors affecting adhesion during the manufacturing process may include (i) chemical bond, (ii) interfacial contamination, (iii) induced modification of an interface, (iv) mechanical interlocking, and (v) stability under working conditions, that is, the effect of high temperature and high humidity.

It should be emphasized that the processes related to the interfaces (e.g., contamination diffusion, reaction between various species, and induced modification) influence the properties and degradation of practical interfaces. In addition, the interface properties (and thus adhesion) may undergo changes with time under working conditions due to temperature and humidity excursions.

The two principal failure mechanisms associated with adhesive bonding are adhesive failure (i.e., the failure at the interface between the adhesive and one of the adherends) and cohesive failure (i.e., the internal breakdown of either the adhesive or one of the adherends). The causes of the failure in adhesively bonded joints may include stresses, temperature, humidity, fatigue, solvents, and variable processing conditions.

A wide variety of characterization techniques are available for failure analysis and quality control of adhesive bonding. Some techniques can be employed *in situ* and others can only be used *ex situ*. Techniques for the analysis of adhesives include thermal analysis techniques such as thermal gravimetric analysis (TGA), differential thermal analysis (DTA), differential scanning calorimetry (DSC), and thermal mechanical analysis (TMA). Various microscopy techniques are available for the analysis of a wide range of materials properties. The microscopy techniques that can provide information related to adhesive bonding include scanning electron microscopy (SEM) and transmission electron microscopy (TEM), scanning acoustic microscopy (SAM), and scanning probe microscopy (SPM).

Raman spectroscopy is one of the most promising methods for the analysis of curing processes related to adhesives, and infrared absorption spectroscopy can be used to determine molecular bonding and coordination.

The surface characterization techniques for adhesives include X-ray photo-electron spectroscopy (XPS), Auger electron spectroscopy (AES), and secondary ion mass spectrometry (SIMS).

Some of the critical issues related to adhesive bonding include the mechanical properties of bonded structures, and some of the most important tests relate to the adhesive's strength, its fracture toughness, and the resistance of the bonded joint to solvents and environmental conditions of high temperature and humidity.

CHAPTER 8

Future Directions and Developments

As noted in the Introduction (Chapter 1), one of the critical issues in the manufacturing of optical assemblies for fiber-optic telecommunications applications is that of automation, involving rapid machine-assisted alignment and attachment processes directed at large-volume manufacturing. This trend toward automation of photonics assembly will continue, since it is highly desirable from both a technical and an economical points of view. This would require the concurrent developments of reliable, in-line, and self-regulated curing methodologies and systems, equipped with real-time cure monitoring systems.

The developments in polymer science are expected to provide new and improved (and tailor-made) adhesives, based on the fact that a wide variety of polymers with designed properties can, in principle, be synthesized for joining different combinations of materials. These developments may eventually lead to adhesives with the following characteristics:

(a) Improved moisture resistance
(b) Low shrinkage
(c) Adhesion to low-surface-energy materials such as rubbers
(d) Adhesives with no requirements for surface treatments
(e) Adhesives with reduced use of solvents and low outgassing
(f) Adhesives that can tolerate harsh environments, such as UV radiation and a corrosive environment, as well as wide temperature fluctuations

Some of the promising developments are related to hybrid adhesives with improved properties (Messler, 2000). These include the so-called *adhesive alloys* (i.e., physical blends of thermoplastic and thermosetting types of adhesive), which behave as both thermoplastic and thermosetting types of adhesives. In such alloys, a homogeneous mixing is realized (but not at a molecular scale), and, thus, the basic properties of each constituent adhesive are preserved. The future developments may also lead to hybrid adhesives with mixing taking place at the molecular scale, resulting in a hybrid system with novel properties. The future applications of adhesives, in addition to being facilitators of mechanical bonding, will also require the development of adhesives with the characteristics that ensure sufficient electronic and optical path integrity at the interfaces, especially materials that are able to withstand the higher power densities of pump lasers. In this context, one should also consider any possible nonlinear effects in the adhesive as the result of the high-power optical beam passing through the assembly.

There is also a continuing effort directed at developing new formulations of adhesives, including those containing (inorganic) nanofillers for various purposes (e.g., controlling such properties as shrinkage, viscosity, electrical conductivity, thermal conductivity). Such nanocomposites also reveal great improvements in such properties as hardness and toughness, as well as scratch and abrasion resistance (Roscher et al., 2002). In this case, monodisperse silica nanoparticles (i.e., SiO_2 with particle size distribution in the range between 10 and 50 nm) are employed in various silica nanocomposites (SNCs). The great advantage of such nanocomposites is the fact that, since the particle sizes are much smaller in relation to the wavelength of the visible light, the transparency of the material is not affected by the incorporation of such nanoparticles.

Other types of adhesives with advantageous features will include adhesives with dual photoinitiators (for surface cure and bulk cure) and delay cure adhesives, which are activated by UV light and will cure within a specified time window to allow for alignment.

It is also highly desirable to develop adhesives that would contain ingredients for providing complete self-functionality in relation to their processing. In such cases, an adhesive would be adaptable to different materials with the capability to facilitate an intervention-free sequence of surface self-treatment, self-activation, wettability, and bonding. Such an adhesive, containing appropriate constituents, would also be self-healing and thus resist degradation. In many respects, this would be a "smart adhesive."

The issue of self-healing of polymer adhesives has been recently addressed in relation to microcracking, which is typically induced by thermal and mechanical fatigue. Such efforts directed at autonomic healing of cracks in a polymeric material are of great importance. As reported recently (White et al., 2001), for autonomic healing, the material incorporates a microencapsulated healing agent, which is released upon crack formation. The contact with an embedded catalyst activates the polymerization of the healing agent, followed by bonding of the crack faces (White et al., 2001). A transparent organic cross-linked polymer material, also reported recently (Chen et al., 2002), can be repeatedly self-repaired, if it develops cracks, by a process of heating and subsequent cooling, which results in covalent bonds between the polymer chains being rearranged during heat treatment. Thus, these covalent bonds facilitate the self-healing characteristics of the polymer. In this case, the bonds between the polymer chains are broken during the heating stage (at temperatures above 120 °C), and these bonds restructure during cooling, resulting in the repair of cracks or fractures (Chen et al., 2002). The advantage of such a material is related to the facts that it does not require any additional components (e.g., a catalyst) and that it can be repeatedly self-repaired employing a relatively easy thermal treatment.

The continuing developments in polymer science and adhesive applications have to be related to other joining techniques, such as laser welding and soldering (see Table 3.1). In the comparison of joining techniques, the main disadvantages of adhesive bonding relate to (i) its susceptibility to moisture and outgassing, (ii) its lack of component adjustment following bonding, (iii) its limited service temperature range (related to the glass transition temperature and/or chemical degradation of the adhesive), and (iv) hermeticity issues. Taking into account the advantages of adhesive bonding, the improvements related to these issues will make adhesive bonding more competitive, compared to laser welding, as a joining technique of choice in a variety of applications. The ultimate question is whether it is feasible to develop adhesives (polymer or otherwise such as organic–inorganic nanocomposite type) with improved mechanical and thermal properties.

Applications of *self-assembly* in relation to adhesion science and technology may also become increasingly important. One of the primary objectives in this context is to facilitate improved adhesion properties. One example is employing silane compounds for attaching a layer of polymeric material to an inorganic substrate.

The continuing trends toward the development of nanoscale characterization techniques, based on scanning probe microscopy methods, would also make it possible to compare experimental observations on the nanoscale with macroscopic properties related to adhesion, so that the scaling behavior of adhesion can be ascertained. This would also aid in revealing the relative importance of the microscopic mechanisms that are responsible for adhesion properties.

Another major issue related to photonics packaging is that of *standardization*. The relevance of the standards concerned with adhesive bonding in fiber-optic manufacturing, for example, can be related to two major perspectives. First, such standards should apply to the physical package of the component; second, the standards should apply to environmental testing of the adhesive that is employed in the assembly process. A useful comparison can be made with the microelectronics industry, where the so-called JEDEC (Joint Electron Device Engineering Council) standards apply to the complete manufacturing process for the package or, in a specific case, the *printed circuit boards* (PCBs). Predetermined standards apply to the physical dimensions of the PCB, interconnects, chips, and other components within or on the package. As a result, highly developed automated assembly lines are employed for efficient manufacturing of the electronics products. Standards that are related to the performance of the component are also established based on the end use of the product. It would be highly desirable to develop similar standards for fiber-optics component manufacturing as well. At present, the component package represents between about 60 and 85% of the total cost. Because of the lack of standards for the physical characteristics of the package, however, the process has been difficult to automate. As a result, manual assembly processes are employed, and the yield and throughput issues, which could be addressed by automation, cannot be realized. From the adhesive standpoint, the two standards that are most relevant to the fiber-optics manufacturer are the G-1209-CORE and G-1221-CORE testing requirements (administered by Telcordia Technologies). These standards have been established to determine if a component meets its prescribed performance characteristics after being exposed to accelerated lifetime testing related to exposure at 85 °C and 85% relative humidity for 2000 h.

The trends toward automation of photonics assembly, requiring the development of self-regulated curing methodologies and systems equipped with real-time cure-monitoring systems, will also be accompanied by the development of a light delivery system configured for specialized applications of curing, including nonflat curing geometries (e.g., geometrically conformal for fiber-optic components and systems). The light-emitting diodes with sufficient radiation intensity may also become the candidates for sources of radiation having tailored spectral content.

The total integration of the adhesive bonding process (including automatic alignment, adhesive dispensing, curing, and cure monitoring) is also highly desirable, since it could provide a reliable and reproducible means of adhesive bonding in a mass production environment.

Glossary

Absorption (In relation to a fiber-optic cable), the loss of power due to conversion of optical power into heat, caused mainly by impurities (e.g., transition metals and hydroxyl ions).

Accelerator (In relation to the cure of adhesives and sealants), a substance used to speed up the cure.

Acrylics Thermoplastic synthetic resins based mainly on acrylic esters with a wide range of performance characteristics (e.g., excellent optical clarity, strength, and durability) and exhibiting fast cure times and excellent adhesion to various substrates. (Note, however, the limited resistance to chemical exposure and elevated temperatures.)

Adherend A material being bonded.

Adhesion The state in which two bodies are held together at an interface by intermolecular forces, including chemical bonding across the interface, or interlocking action (using rough surfaces) or both.

Adhesion promoter A coating additive (also called bonding agent) to improve adhesion by either surface pretreatment of the substrate or as a component in a formulation producing improved substrate wetting and formation of chemical bonds across the interface.

Adhesive A material capable of holding two other materials together in a practical manner by surface attachment that resists separation.

Adhesive bonding The assembly of materials by employing adhesives; joining of two materials by employing adhesives.

Adhesive failure Type of breakage of an adhesive bond, with the separation occurring at the interface between the adhesive and the adherend.

Adhesive strength (Also referred to as bond strength), the strength with which two surfaces are held together with an adhesive.

Amorphous Having no ordered arrangement. Polymers are referred to as amorphous when their chains are tangled up in various ways. (Polymers are referred to as crystalline when their chains are lined up in an ordered manner.)

Amplitude modulation   Transmission technique in which the information is encoded in the amplitude of a carrier.

Analog   Signals that are continually varying, as compared to signals that are digitally encoded.

Angular tilt   An angle produced by the axes of two fibers that are being joined; this results in an extrinsic loss that relates to the joining method and/or hardware.

Anisotropic   Having physical and/or mechanical properties varying with direction relative to reference axes intrinsic to the material.

Attenuation   The decrease in optical power, that is, in amount of radiant energy per unit time, during passage through a fiber-optic cable. Attenuation is due to such processes as absorption and scattering, and it is typically measured in decibels per kilometer, that is, dB/km. (Note that calculations and equations related to attenuation include the minus sign.)

Attenuator   (In relation to a fiber-optic cable), a passive optical component that reduces the optical power propagating in a fiber.

Axial ray   A light ray traveling along the axis of an optical fiber.

Backscattering   The return of a portion of scattered light in the direction that is generally opposite to that of the original direction of propagation.

Bandwidth   (In relation to a fiber-optic cable), the measure of information-carrying capacity of a fiber-optic cable.

Bend loss   (In relation to a fiber-optic cable), the loss resulting from the fiber-optic cable being curved through a restrictive radius or curvature.

Bend radius   (In relation to a fiber-optic cable), the radius corresponding to the value that a fiber-optic cable can bend prior to breakage or increase in attenuation.

Bit   A binary digit (typically either 0 or 1), corresponding to the smallest representation of information in computing and communications systems.

Bond line   A layer of adhesive that joins two adherends.

Buffer   A protective layer over the fiber-optic cable; this may include a coating, a jacket, or a hard tube.

Bundle   Numerous individual fiber-optic cables within a single jacket.

Byte   A unit composed of 8 bits.

Cable   Same as a fiber-optic cable, that is, an assembly of optical fibers and additional material that facilitates environmental protection and optical insulation of the inner optical wave guide.

Cable assembly   A fiber-optic cable having connectors installed on one or both ends. In the case of a connector being attached to one end of the cable only, it is referred to as a *pigtail*; in the case of a connector being attached to both ends, it is referred to as a *jumper*.

Catalyst A chemical substance that initiates or alters the rate of chemical reaction. (Note that the catalyst itself is not affected by the reaction, and it is typically incorporated as a component in a two-part thermosetting adhesive.)

Chromatic dispersion Spreading of a light pulse due to the difference in refractive indices at different wavelengths. This is a result of the facts that (i) the speed of an optical pulse propagating down the fiber varies with wavelength, and (ii) in a practical light source, the presence of components at different wavelengths leads to a pulse broadening.

Cladding Glass or plastic material, which surrounds the core of the optical fiber, having a lower refractive index as compared with a core material of higher refractive index. Thus, optical cladding facilitates total internal reflection for the propagation of light in the optical fiber.

Cladding mode (In relation to an optical fiber), a mode that is confined to the cladding or a lightwave that propagates in the cladding. (Note that this mode is eliminated after propagating a short distance, on the order of a meter, due to high attenuation in the cladding.)

Cleaving The breaking of an optical fiber that ensures its smooth end surface.

Coating (In relation to an optical fiber), a material deposited on a fiber-optic cable (during the drawing process) in order to protect it from the environment.

Cohesion The state in which the constituents of a material are held together by chemical and physical forces.

Cohesive failure Type of breakage of an adhesive bond, with the separation occurring within the adhesive bond layer.

Connector A mechanical device that is attached to the end of a fiber-optic cable, or light source, or receiver, which connects to a similar device. This facilitates optical coupling of light into and out of a fiber-optic cable, with a possibility of repeatable connection and disconnection of the fiber-optic cable from a device.

Contaminant A foreign substance or an impurity that influences properties of the material, including adhesion properties.

Core The central region (about the longitudinal axis of an optical fiber) that supports guiding of the optical signal. To ensure the guiding of the optical signal, the refractive index of the core must be greater than that of the surrounding cladding.

Core eccentricity (In relation to an optical fiber), a measure of the displacement of the core center relative to the cladding center.

Coupler A multiport device that is employed for the distribution of optical power.

Coupling efficiency The efficiency of optical power transfer between two optical components.

Coupling loss The loss in optical power due to coupling light from one optical device to another. Typical cases are due to intrinsic losses (i.e., nonideal fiber parameters) and extrinsic losses (i.e., mechanical effects).

Creep A time-dependent component of strain resulting from an applied stress. (Plastic deformation resulting from a constant load over a prolonged period at relatively high temperatures.) In practice, creep may lead to dimensional change, occurring with time in the presence of fast deformations (e.g., repeated cycling). (Note that creep at ambient temperatures is also referred to as cold flow.)

Critical angle (In relation to the total internal reflection in an optical fiber), the smallest angle at which a light ray will be totally reflected within the fiber and thus guided down the fiber.

Cross-linking Formation of a massive molecule by linking individual polymer chains together by covalent bonds.

Crystal Molecules arranged in an ordered manner. In polymer crystals, the chains are lined up in order and are kept bound together by secondary interactions.

Cure A process of conversion of an adhesive from a liquid to a solid state, accompanied by a physical or chemical modification of the adhesive. Converting a liquid adhesive into a solid state can be accomplished, for example, by light (photoinduced) curing or polymerization.

Cure stress A residual internal stress originating between different components (typically due to different coefficients of thermal expansion) during the curing process.

Cyanoacrylates Adhesives with very fast curing characteristics and with the capability of quickly bonding to various metallic and nonmetallic substrates.

Debond An intentional separation of a bonded joint (this is typically performed for rework purposes).

Decibel (dB) (In fiber-optic terminology), a measure of loss (or attenuation), computed as a standard logarithmic unit for the ratio of two powers, defined as $dB = 10 \log(P_1/P_2)$.

Degradation A detrimental modification of physical properties, chemical structure, and/or appearance.

Degree of polymerization The average number of monomer units in a polymer chain molecule.

Demultiplex (In relation to a fiber-optic cable), the process of separating optical channels that have been multiplexed in order to share a common transmission medium.

Detector (In relation to the detection of optical radiation), a transducer that converts incident optical energy to an electrical signal.

Digital A data format employing a discrete (finite) number of levels for transmission of information (the binary format is a specific case corresponding to two levels).

Dispersion   A term describing the phenomena that result in a broadening or spreading of propagating light through an optical fiber.

Ductility   The amount of plastic strain that a material can tolerate prior to fracture.

Elastomer   A polymeric material that can be stretched (at ambient temperatures) as a minimum to twice its initial length by a deforming force (note that with the removal of the deforming force, the elastomer returns to its initial length).

End-to-end loss   (In relation to a fiber-optic cable data link path), the optical loss consisting of the loss due to the fiber-optic cable, splices, and connectors.

Epoxy   An adaptable group of thermosetting polymers for various applications (adhesion, sealing, encapsulation, and coating), which can be either two-part (room-temperature curing) or one-part (heat curing) and have great physical strength and excellent resistance to environmental damage and superior dimensional stability with excellent service temperature range. (Epoxies are typically employed as structural adhesives and as electrical insulation materials, although specific formulations also have high electrical and thermal conductivities.)

Epoxy resin   A resin that contains an epoxy (or epoxide) group, which can be connected in its final configuration by a chemical reaction.

Erbium-doped fiber amplifier (EDFA)   A type of fiber-optic cable that amplifies 1550-nm optical signals when pumped with a light source in the wavelength range between 980 and 1480 nm.

Exotherm   The amount of heat released; the temperature versus time curve of a chemical reaction or a phase change giving off heat.

Exothermic   Type of chemical reaction that releases heat (as opposed to an endothermic reaction, which requires heat to continue).

Fatigue   The failure or degradation of mechanical properties following repeated applications of stress. Fatigue tests provide data related to the ability of a material to resist the growth of cracks, which ultimately cause failure due to the large number of cycles.

Ferrule   The component (typically cylindrical in shape with a grip through the center) of a connector that confines the stripped end of a fiber-optic cable and facilitates its alignment.

Fiber   (In relation to photonics), a dielectric material that guides light (i.e., a wave guide). *See* **Optical fiber**.

Fiber-optic attenuator   The component of a fiber-optic transmission system that diminishes the optical signal power in order, for example, to control the optical power within the limits of the optical receiver.

Fiber-optic cable   A cable including one or a number of optical fibers.

Fiber-optic communication system   The transfer of optical energy (modulated or unmodulated) through an optical fiber system.

Fiber-optic connector   A rapid connect/disconnect assembly for interconnecting (i) a light source to a fiber-optic cable, (ii) a fiber-optic cable to another fiber-optic cable, or (iii) a fiber-optic cable to a light detector.

Fiber-optic coupler   A component for interconnecting several fiber-optic cables in a bidirectional system.

Fiber optics   The branch of photonics related to the transmission of light through optical fibers for the purpose of data transmission (using pulses) or the transmission of images (employing optical fiber bundles).

Fillers   (In relation to adhesive materials), a relatively inert substance added to an adhesive material for modifying its properties (e.g., improving ease of application, strength, dimensional stability, durability, or hardness).

Fillet   (In relation to an adhesive joint), the segment of an adhesive that fills the corner formed at the joint between two adherends.

Gap loss   (In an optical fiber), the optical power loss due to the space between axially aligned fiber-optic cables.

Glass transition temperature (T_g)   The approximate midpoint of the temperature range corresponding to the region where the glass transition occurs. T_g represents the transition temperature between the glassy and viscoelastic states of polymers. Below T_g, polymers become rigid (and, in some cases, brittle); above T_g, polymers are rubber-like (up to the flow temperature T_f).

Graded-index (GRIN) optical fiber   An optical fiber in which the refractive index of the core varies with the core radius such that the refractive index gradually decreases away from the center of the core to the outer radius. Such an optical fiber serves the purpose of reducing modal dispersion.

Green strength   A measure of the capability of an adhesive to carry a load during the green time.

Green time   The time between dispensing and solidification of an adhesive. (During this time window, the positioning of parts is still realizable.)

Hardener   A substance that is added to an adhesive composition in order to control (promote) the curing reaction by taking part in it.

Index-matching material   (In relation to an optical interconnection), a material that has a refractive index similar to that of the fiber core.

Index of refraction   *See* **Refractive index (*n*).**

Index profile   (In a graded-index optical fiber), the refractive index of the core varies with the core radius such that the refractive index gradually decreases away from the cylindrical axis.

Infrared (IR)   The region of the electromagnetic spectrum in the range between about 750 nm and 1 mm.

Inhibitor   A substance that is added to an adhesive composition in order to slow down the rate of chemical reaction.

Insertion loss   The loss of power due to the insertion of a component (e.g., a connector or splice) into a previously continuous path.

Joint (or adhesive joint) The section at which two adherends are held together by a layer of adhesive.

Lap joint A joint prepared by placing one adherend partly over another adherend and bonding together the overlapped segments.

Laser (Acronym for light amplification by stimulated emission of radiation), a light source generating coherent and near-monochromatic light by employing stimulated emission of radiation. (Typically, solid-state semiconductor lasers are employed in fiber-optic communication systems.)

Lateral displacement loss The loss of power due to lateral displacement from optimum alignment between two fiber-optic cables or between a fiber-optic cable and an active device.

Light (In relation to photonics), the region of the electromagnetic spectrum in the range between the near-ultraviolet region (about 300 nm) and the infrared region (to about 30 μm).

Light-emitting diode (LED) A semiconductor device (diode) that emits light (spontaneously) from the p–n junction under an applied forward current.

Light (induced) degradation (photodegradation) Process related to chemical decomposition and structural changes in polymers, in the presence of light, that results in both polymer damage and cross-linking; this is followed by the formation of molecular fragments and a highly cross-linked system, which consequently may lead to internal stress, cracking, and loss of adhesion. (To avoid such degradation, light absorbers are employed.)

Loss (In an optical fiber), optical signal attenuation measured in decibels.

Methacrylates A group of thermoplastics having excellent optical clarity, abrasion resistance, and good physical strength.

Microbend loss The loss of power due to microscopic bends in a fiber-optic cable.

Misalignment loss The loss of power due to angular misalignment, lateral displacement, or end separation.

Modal dispersion The dispersion due to the time difference for different rays traversing a fiber-optic cable.

Modifier Any inert chemical ingredient that is added to an adhesive composition in order to modify its properties.

Modulation The process related to the characteristic of one wave (i.e., the carrier) being modified by another wave (i.e., the information signal). The types of modulation are, for example, amplitude modulation (AM) and frequency modulation (FM).

Modulus of elasticity The ratio of the stress (or load applied) to the strain (or deformation) produced in a material that is elastically deformed. (Often referred to as Young's modulus.)

Monomer A simple molecule with reactivity to make possible the formation of a polymer by the linking of many such simple molecules.

Multimode fiber   Optical fiber that supports more than one propagating mode.

Multiplexing   (In relation to an optical fiber), the transmission of two or more signals through a single communications channel (e.g., time division multiplexing and wavelength division multiplexing).

NA mismatch loss   The loss of power at a joint that occurs when the transmitting half has an NA greater than the NA of the receiving half. The loss occurs when coupling light from a light source to a fiber-optic cable, from a fiber-optic cable to a fiber-optic cable, or from a fiber-optic cable to a detector.

Numeric aperture (NA)   (In relation to an optical fiber), the ability of an optical fiber to collect light, described by the maximum angle to the axis of an optical fiber at which light will be received and propagated through it.

Oligomer   A polymer that is composed of a few monomer units.

Optical cable   An assembly of fiber-optic cable and other materials that provides environmental and mechanical protection.

Optical fiber (optical wave guide, or light guide)   A filament of transparent dielectric material (glass or plastic) that is typically circular in cross section and that guides light. In general, an optical fiber consists of a cylindrical *core* (with higher refractive index) surrounded by (in intimate contact) a *cladding* of similar geometry and lower refractive index, so that the light can be guided by the fiber.

Optical isolator   A component that is employed for blocking out reflected and other undesirable light.

Optical window   Wavelength range of a fiber-optic cable corresponding to a very low attenuation. For example, fiber-optic data communication systems employing LED sources operate in windows at 850 or 1300 nm, and those employing laser sources operate in windows at 1310 or 1550 nm.

Optoelectronics   Instrumentation for the generation of light (e.g., lasers and light-emitting diodes), amplification of light, control of light, and detection of light. In other words, (i) the device functions as an electrical-to-optical transducer or optical-to-electrical transducer; (ii) the device operation requires electrical energy and depends on sensing and controlling this energy; and (iii) it is related to a device that responds to optical power, emits or changes optical radiation, or makes use of optical radiation for its operation.

Peel strength   The resistance of an adhesive to be stripped from a bonded joint.

Photodetector (photodiode)   A semiconductor device (an optoelectronic transducer) that produces a photocurrent in response to absorbed incident optical power; it is used as a detector in a fiber-optic cable data link.

Photoinitiator   An additive that is capable of absorbing the light energy from an external source and transferring it to monomer molecules, resulting in the generation of active centers for polymerization.

Photonics   The technology field related to the generation, transmission, and control of light for communications and information processing. It includes the emission, deflection, amplification, transmission, and detection of light by employing various optical components and devices, such as fiber optics, light sources (e.g., lasers), and optoelectronic instrumentation.

Pigtail   A relatively short segment of optical fiber that is permanently attached to a component (e.g., a source, coupler, connector, or detector) and is employed for coupling of power between it and the transmission fiber.

Pistoning   The axial movement of an optical fiber in and out of a ferrule end as a result of, for example, temperature variations.

Plastic   A material containing organic polymers of large molecular weight; it is solid in its final form, and it can be shaped by flow during the processing or manufacturing stage.

Plastic-clad silica fiber-optic cable   A fiber-optic cable including a plastic cladding and a glass core.

Plastic fiber-optic cable   A fiber-optic cable including a plastic cladding and a plastic core.

Plasticizer   A component added to an adhesive composition in order to enhance flow, flexibility, and deformation. (Note that the addition of a plasticizer may also reduce tensile strength and elastic modulus and increase toughness and impact strength.)

Poisson's ratio   The ratio of the change in lateral width per unit width to the change in axial length per unit length, caused by the axial stressing or stretching of a material.

Polymer   An organic material; macromolecules formed by the linking of many simpler molecules known as monomers. Polymers are formed by polymerization (addition polymer) or polycondensation (condensation polymer). When two or more monomers are involved, the material is referred to as a copolymer.

Polymerization   A chemical reaction resulting in the linking of molecules of monomers and the forming of polymers through chain growth (i.e., addition polymerization) or step growth (i.e., condensation polymerizations). When two or more monomers are involved, the process is referred to as copolymerization.

Polyurethanes   A group of polymers with widely varying properties. They may be thermoplastic or thermosetting, rigid or flexible, and the properties of any of these types can be modified within wide limits to meet the requirements of a specific application.

Postcure   An additional treatment (typically relating to heat treatment) applied to an adhesive assembly following the initial cure; such treatment is directed at modifying specific properties (e.g., increasing the glass transition temperature) or completing the cure.

Pot life   The time period during which a reacting composition continues to be fit for its planned processing use following mixing with a reaction initiating agent, such as an accelerator or catalyst, or exposure to curing conditions.

Potting   The process of filling a space, including, for example, electronic components, for protection against vibration or environmental exposure (e.g., moisture, chemicals, heat).

Primer   A coating that is applied to the surface of a material (in order to improve the performance of the adhesive bond) before the application of an adhesive.

Rayleigh scattering   The scattering of light, resulting in optical power losses, that is due to small inhomogeneities in material density or composition. The Rayleigh scattering follows the λ^{-4} dependence, and it sets a theoretical lower limit to the attenuation of propagating light as a function of wavelength, varying from about 10 dB/km (at 0.5 μm) to 1 dB/km (at 0.95 μm).

Receiver   (In relation to a fiber-optic communications link), an electronic package that converts optical signals to electrical signals.

Refraction   The bending of a beam of light at an interface between two dissimilar media or a single medium (i.e., a graded-index medium) having refractive index that is a continuous function of position.

Refractive index (n)   The ratio of the velocity of propagation of an electromagnetic wave in a vacuum to its velocity in the medium.

Resin   An organic material exhibiting a propensity to flow with applied stress. It is typically employed as an adhesive or as a matrix to contain the reinforcement in a composite material; the organic matrix may be a thermoplastic or a thermoset and may also include a variety of additives.

Shear   An action or stress due to applied forces that results in two adjacent sections of a material sliding relative to each other in a direction that is parallel to their plane of contact.

Shear strength   The ability of a plastic material to endure shear stresses.

Silicones   Polymeric materials containing the main polymer chain that consists of alternating silicon and oxygen atoms and carbon-containing side groups.

Single-mode fiber   An optical fiber with a relatively small core (a diameter of about 10 μm) that supports only one mode of light propagation above the cutoff wavelength. Since the dispersion and power loss through such a fiber are relatively low, it is suitable for long-distance transmission.

Snell's law   The law that describes the bending (refraction) of light rays crossing the boundary between dissimilar media (materials); in effect, the refraction is determined by the relative refraction indices of the two media (materials).

Splicing   A permanent joining method (without a connector) of the ends of similar optical fibers. Examples of the joining method are thermal fusing or mechanical. (Thermal fusing is realized by the application of sufficient localized heat for melting or fusing the ends of the optical fiber cables, resulting in a single continuous optical fiber cable.)

Star coupler   A type of coupler (for an optical fiber) in which power at any input port is distributed to all output ports.

Step-index optical fiber   An optical fiber in which the refractive indices of both the core and the cladding (note again that the refractive index of the core is always higher than that of the cladding) are uniform across their respective cross sections, so that a sharp step in the refractive index profile is established at the interface between the core and the cladding.

Stiffness   The relationship of load and deformation; a term often used when the relationship of stress to strain does not conform to the definition of Young's modulus.

Strain   The elongation per unit length of a material; the unit change (due to stress) in the size or shape of a material in relation to its original size or shape.

Stress   The force exerted per unit area at a point within a plane.

Substrate   (In relation to adhesives), a material to which an adhesive is applied.

Surface tension   The force, present in a liquid–vapor phase interface, that tends to reduce the area of the interface.

Surfactant   (Surface active agent), an additive that reduces surface tension and, thus, controls surface-related processes (e.g., improves substrate wetting) in various applications.

Tack   The property of an adhesive to form a sufficient bond instantly following the establishment of direct contact between the adhesive and the adherend by using relatively low pressure.

Tensile strength   The maximum stress that a material can be exposed to without tearing during stretching under tensile load.

Thermoplastics   Plastic materials (such as PVC, polystyrene, nylon, polycarbonates) that can repeatedly become elastic or melt when heated (and, thus, can be shaped by flow into specific objects by molding or extrusion) and return to their hardened state by cooling through a temperature range characteristic of the specific material.

Thermosets   Polymeric materials (such as polyesters, acrylics, epoxies, and phenolics) that undergo an irreversible chemical cross-linking reaction going from liquid to solid.

Total internal reflection   Total reflection (back into a material) of light that strikes the interface of another material (with a lower refractive index) at an angle below the critical angle.

Toughness   A measure of the ability of a material to absorb energy.

Tree coupler   A passive photonics component that facilitates the distribution of power from one-input to more than two-output fiber-optic cables.

Two-component adhesive   An adhesive provided in two parts that are mixed before application.

Ultraviolet (UV)   The region of the electromagnetic spectrum in the range between about several nanometers to about 400 nm (i.e., with the longest wavelength being just below the visible spectrum).

UV curing adhesive   An adhesive in which the chemical curing process is initiated by ultraviolet (UV) light.

Viscoelasticity   A material's property related to a combined elastic and viscous behavior.

Viscosity   The property of resistance to flow exhibited by a material.

Wave guide   A two-dimensional substrate that carries light in channels etched in the material.

Wavelength division multiplexer   A passive device employed to separate optical signals of different wavelengths transmitted in a single fiber-optic cable.

Wavelength division multiplexing (WDM)   Simultaneous transmission of several optical signals having different wavelengths in the same fiber-optic cable; this enables several different communications channels to share the same fiber-optic cable.

Wetting   The process of spreading a liquid on or into a solid surface. Typically, the *spontaneous wetting* process is driven by minimization of the free surface energy, whereas the *forced wetting* process necessitates the use of an external force (note that the removal of that force results in de-wetting).

Bibliography

RECENT BOOKS ON GENERAL TOPICS RELATED TO ADHESIVES AND ADHESION

A. J. Kinloch, "Adhesion and Adhesives: Science and Technology," Chapman & Hall, New York, 1987.

J. N. Israelachvili, "Intermolecular and Surface Forces," Academic, New York, 1991.

L. H. Lee, Ed., "Adhesive Bonding," Plenum, New York, 1991.

E. W. Flick, "Adhesives, Sealants and Coatings for the Electronics Industry," Noyes, Park Ridge, NJ, 1992.

S. P. Pappas, Ed., "Radiation Curing: Science and Technology," Plenum, New York, 1992.

R. W. Messler, Jr., "Joining of Advanced Materials," Butterworth–Heinemann, Stoneham, MA, 1993.

K. L. Mittal, Ed., "Contact Angle, Wettability and Adhesion," VSP, Utrecht, 1993.

A. Pizzi and K. L. Mittal, Eds., "Handbook of Adhesive Technology," Dekker, New York, 1994.

J. P. Fouassier, "Photoinitiation, Photopolymerization, and Photocuring: Fundamentals and Applications," Hanser Verlag, New York, 1995.

I. Benedek and L. J. Heymans, "Pressure-Sensitive Adhesives Technology," Dekker, New York, 1997.

J. Comyn, "Adhesion Science," Royal Society of Chemistry, Cambridge, UK, 1997.

A. V. Pocius, "Adhesion and Adhesives Technology: An Introduction," Hanser Verlag, New York, 1997.

W. J. Van Ooij and H. R. Anderson, Jr., Eds., "First International Congress on Adhesion Science and Technology," VSP, Utrecht, 1998.

L. P. DeMejo, D. S. Rimai, and L. H. Sharpe, Eds., "Fundamentals of Adhesion and Interfaces," Gordon & Breach, New York, 1999.

J. Liu, "Conductive Adhesives for Electronic Packaging," Electrochemical Publications, Port Erin, Isle of Man, UK, 1999.

K. L. Mittal and A. Pizzi, Eds., "Adhesion Promotion Techniques: Technological Applications," Dekker, New York, 1999.

Adhesive Bonding in Photonics Assembly and Packaging
© 2003 by American Scientific Publishers

K. L. Mittal, Ed., "Polymer Surface Modification: Relevance to Adhesion," Vol. 2, VSP, Utrecht, 2000.

E. M. Petrie, "Handbook of Adhesives and Sealants," McGraw–Hill, New York, 2000.

G. Gierenz, Ed., "Adhesives and Adhesive Tapes," Wiley–VCH, Weinheim/New York, 2001.

K. Kendall, "Molecular Adhesion and Its Applications," Kluwer Academic/Plenum, New York, 2001.

RECENT REVIEWS ON ADHESIVES AND ADHESION

U. Landman, W. D. Luedtke, N. A. Burnham, and R. J. Colton, *Science* 248, 454 (1990).

P. Attard and J. L. Parker, *Phys. Rev. A* 46, 7959 (1992).

D. E. Packham, *Int. J. Adhesion Adhesives* 16, 121 (1996).

M. Vergeles, A. Maritan, J. Koplik, and J. R. Banavar, *Phys. Rev. E* 56, 2626 (1997).

G. V. Dedkov, *Phys. Status Solidi A* 179, 3 (2000).

R. W. Messler, Jr., *Assembly Automation* 20, 118 (2000).

H. Rafii-Tabar, *Phys. Rep.* 325, 239 (2000).

References

Adams, R. D., and Wake, W. C., "Structural Adhesive Joints in Engineering," Elsevier, London, 1984.

Adhihetty, I., Hay, J., Wei, C., and Padmanabhan, P., in "Fundamentals of Nanoindentation and Nanotribology" (N. R. Moody, W. W. Gerberich, N. Burnham, and S. P. Baker, Eds.), Vol. 522, p. 317. Materials Research Society, Warrendale, PA, 1998.

Adolf, D., and Martin, J. E., *J. Compos. Mater.* 30, 13 (1996).

Bertrand, N., Drevillon, B., Gheorghiu, A., Senemaud, C., Martinu, L., and Klemberg-Sapieha, J. E., *J. Vac. Sci. Technol., A* 16, 6 (1998).

Bourne, P. A., and Thadani, V., *Proc. SPIE* 4771 (2002).

Brewis, D. M., and Briggs, D., Eds., "Industrial Adhesion Problems," Orbital Press, Oxford, 1985.

Brundle, C. R., Evans, C. A., and Wilson, S., Eds., "Encyclopedia of Materials Characterization," Butterworth–Heinemann, Boston, 1992.

Burnham, N. A., and Colton, R. J., *J. Vac. Sci. Technol., A* 7, 2906 (1989).

Chen, X., Dam, M. A., Ono, K., Mal, A., Shen, H., Nutt, S. R., Sheran, K., and Wudl, F., *Science* 295 1698 (2002).

Cheng, L., Scriven, L. E., and Gerberich, W. W., in "Fundamentals of Nanoindentation and Nanotribology" (N. R. Moody, W. W. Gerberich, N. Burnham, and S. P. Baker, Eds.), Vol. 522, p. 193. Materials Research Society, Warrendale, PA, 1998.

Cheng, L., Xia, X., Yu, W., Scriven, L. E., and Gerberich, W. W., *J. Polym. Sci., Part B: Polym. Phys.* 38, 10 (2000).

Colton, R. J., Barger, W. R., Baselt, D. R., Corcoran, S. G., Koleske, D. D., and Lee, G. U., in "First International Congress on Adhesion Science and Technology" (W. J. Van Ooij and H. R. Anderson, Jr., Eds.), p. 21. VSP, Utrecht, 1998.

Daly, J. G., "Photonics West 2002," 2002.

Darmody, M., and Chadwick, G., "Third Annual Expert Systems in Government Conference Proceedings," p. 145. IEEE, Washington, DC, 1987.

Darmody, M. P., and Schneemann, G. K., "Proceedings of the Second International Conference on Industrial and Engineering Applications of Artificial Intelligence and Expert Systems," Vol. 2, p. 672. ACM, New York, 1989.

Decker, C., *Proc. SPIE* 1279, 50 (1990).

Decker, C., *Nucl. Instrum. Methods Phys. Res., Sect. B* 151, 22 (1999).

Derebail, A., Srihari, K., and Emerson, C. R., *Int. J. Adv. Manuf. Technol.* 9, 93 (1994).

Dürig, U., and Stalder, A., in "First International Congress on Adhesion Science and Technology" (W. J. Van Ooij and H. R. Anderson, Jr., Eds.), p. 109. VSP, Utrecht, 1998.

Ehlers, H., Biletzke, M., Kuhlow, B., Przyrembel, G., and Fischer, U. H. P., *Opt. Fiber Technol.* 6, 344 (2000).

Eldada, L., and Shacklette, L. W., *IEEE J. Selected Topics Quantum Electron.* 6, 54 (2000).

Elliott, J. E., and Bowman, C. N., *Macromolecules* 32, 8621 (1999).

Estes, R. H., "Proceedings of the 1986 International Symposium on Microelectronics," p. 642. International Society of Hybrid Microelectron, Reston, VA, 1986.

Estes, R. H., *Hybrid Circuit Technol.* 8, 44 (1991).

Flick, E. W., "Adhesives, Sealants and Coatings for the Electronics Industry," Noyes, Park Ridge, NJ, 1992.

Fouassier, J. P., "Photoinitiation, Photopolymerization, and Photocuring: Fundamentals and Applications," Hanser Verlag, New York, 1995.

Fouassier, J. P., and Rabek, J. F., Eds., "Radiation Curing in Polymer Science and Technology," Vol. 3, Chap. 7, pp. 219–268. Elsevier, New York, 1993.

Fujisawa, S., Kishi, E., Sugawara, Y., and Morita, S., *Phys. Rev. B* 51, 7849 (1995).

Goodner, M. D., and Bowman, C. N., *Macromolecules* 32, 6552 (1999).

Grau, P., Meinhard, H., and Mosch, S., in "Fundamentals of Nanoindentation and Nanotribology" (N. R. Moody, W. W. Gerberich, N. Burnham, and S. P. Baker, Eds.), Vol. 522, p. 153. Materials Research Society, Warrendale, PA, 1998.

Hubert, M., *Adhesives and Sealants Industry* 28 (2001).

Hudson, A. J., Martin, S. C., Hubert, M., and Spelt, J. K., *ASME J. Electron. Packaging*, (2002).

Imanaka, M., Fujinami, A., and Suzuki, Y., *J. Mater. Sci.* 35, 2481 (2000).

Israelachvili, J. N., "Intermolecular and Surface Forces," Academic, New York, 1991.

Jiang, H., Chou, B., and Beilin, S., "Proceedings of the 1998 International Conference on Multichip Modules and High Density Packaging," p. 7. IEEE, New York, 1998.

Landman, U., Luedtke, W. D., Burnham, N. A., and Colton, R. J., *Science* 248, 454 (1990).

Lee, L. H., Ed., "Adhesives, Sealants, and Coatings for Space and Harsh Environments," Plenum, New York, 1988.

Lin, Y., Liu, W., and Shi, F. G., "Proceedings of ECTC-2002."

Martin, S., and Hubert, M., "The Second IEEE Optoelectronic Packaging Workshop," 2000.

Martin, S., and Hubert, M., "Proceedings of the Technical Program of Microsystems Conference," 2002a.

Martin, S., and Hubert, M., *Proc. SPIE* 4639, 2002b.

McKendry, R. et al. Jpn. J. Appl. Phys. Part 1, 38, 3901 (1999).

Messler, R. W., Jr., *Assembly Automation* 20, 118 (2000).

Mittal, K. L., *Electrocomponent Sci. Technol.* 3, 21 (1976).

Mittal, K. L., Ed., "Polymer Surface Modification: Relevance to Adhesion," VSP, Utrecht, 1996.

Mittal, K. L., Ed., "Polymer Surface Modification: Relevance to Adhesion," Vol. 2, VSP, Utrecht, 2000.

Neumann, A. W., and Spelt, J. K., "Applied Surface Thermodynamics," Dekker, New York, 1996.

Packham, D. E., *Int. J. Adhesion Adhesives* 16, 121 (1996).

Pappas, S. P., Ed., "Radiation Curing: Science and Technology," Plenum, New York, 1992.

Petrie, E. M., "Handbook of Adhesives and Sealants," McGraw–Hill, New York, 2000.

Pignataro, S., in "First International Congress on Adhesion Science and Technology" (W. J. Van Ooij and H. R. Anderson, Jr., Eds.), p. 147. VSP, Utrecht, 1998.

Pocius, A. V., "Adhesion and Adhesives Technology: An Introduction," Chap. 3. Hanser Verlag, New York, 1997.

Rebouillat, S., Letellier, B., and Steffenino, B., *Int. J. Adhesion Adhesives* 19, 303 (1999).

Roscher, C., Adam, J., Eger, C., and Pyrlik, M., "Technical Conference Proceedings of RadTech 2002," p. 321. RadTech International North America, Chevy Chase, MD, 2002.

Sanchez, J. M. et al., *Acta Mater.* 47, 4405 (1999).

Schawe, J. E. K., *Thermochim. Acta* 260, 1 (1995).

Scherzer, T., and Decker, U., *Vib. Spectrosc.* 19, 385 (1999).

Schulz, W. L., Udd, E., Seim, J. M., Perez, I. M., and Trego, A., *Proc. SPIE* 3991 (2000).

Senturia, S. D., *Int. J. Microelectron. Packaging Mater. Technol.* 1, 43 (1995).

Serry, F. M., Strausser, Y. E., Elings, J., Magonov, S., Thornton, J., and Ge, L., *Surf. Eng.* 15, 285 (1999).

Somorjai, G. A., in "First International Congress on Adhesion Science and Technology" (W. J. Van Ooij and H. R. Anderson, Jr., Eds.), p. 3. VSP, Utrecht, 1998.

Strojny, A., and Gerberich, W. W., in "Fundamentals of Nanoindentation and Nanotribology" (N. R. Moody, W. W. Gerberich, N. Burnham, and S. P. Baker, Eds.), Vol. 522, p. 159. Materials Research Society, Warrendale, PA, 1998.

Su, Y. Y., Srihari, K., and Adriance, J., *Comput. Ind. Eng.* 25, 111 (1993).

Suhir, E., "Structural Analysis in Microelectronic and Fiber Optic Systems," Van Nostrand Reinhold, New York, 1991.

Suhir, E., *Future Circuits* 5 (1999).

Suhir, E., *J. Appl. Phys.* 88, 2363 (2000).

Suhir, E., *J. Appl. Phys.* 89, 120 (2001).

Swanson, D. W., and Enlow, L. R., *Proc. SPIE* 3906, 246 (1999).

Wagner, M., Wagner, T., Carroll, D. L., Marien, J., Bonnell, D. A., and Ruhle, M., *MRS Bull.* 22, 42 (1997).

Wesner, D. A., Horn, H., Weichenhain, R., Pfleging, W., and Kreutz, E. W., *J. Adhes. Sci. Technol.* 11, 1229 (1997).

White, S. R., Sottos, N. R., Guebelle, P. H., Moore, J. S., Kessler, M. R., Sriram, S. R., Brown, E. N., and Viswanathan, S., *Nature* 409, 794 (2001).

Wiesendanger, R., "Scanning Probe Microscopy and Spectroscopy," Cambridge Univ. Press, Cambridge, UK, 1994.

Woods, H. F., *Proc. SPIE* 1999, 59 (1993).

Yang, Q. D., and Thouless, M. D., *Int. J. Fract.* 110, 175 (2001).

Xu, Q., Powers, G., and Fisher, Y., "Proceedings of the Technical Program of Microsystems Conference," 2002.

Zhou, S. H., Tang, Z. R., Lin, Y. M., Liu, W. N., Mondal, S. K., and Shi, F. G., *Advanced Packaging* 25 (2002).

Index